IMAGES OF ASIA

The Birds of Peninsular Malaysia

Titles in the series

Aceh: Art and Culture
HOLLY S. SMITH

At the South-East Asian Table
ALICE YEN HO

Betel Chewing Traditions in South-East Asia
DAWN F. ROONEY

The Birds of Java and Bali
DEREK HOLMES and STEPHEN NASH

The Birds of Peninsular Malaysia
DEREK HOLMES and KAREN PHILLIPPS

The Birds of Singapore
CLIVE BRIFFETT and SUTARI BIN SUPARI

The Birds of Sulawesi
DEREK HOLMES and KAREN PHILLIPPS

Borobudur
JACQUES DUMARÇAY

Burmese Dance and Theatre
NOEL F. SINGER

Burmese Puppets
NOEL F. SINGER

A Garden of Eden: Plant Life in South-East Asia
WENDY VEEVERS-CARTER

Gardens and Parks of Singapore
VÉRONIQUE SANSON

The House in South-East Asia
JACQUES DUMARÇAY

Imperial Belvederes: The Hill Stations of Malaya
S. ROBERT AIKEN

The Kris
EDWARD FREY

Lao Textiles and Traditions
MARY F. CONNORS

Life in the Javanese Kraton
AART VAN BEEK

Musical Instruments of South-East Asia
ERIC TAYLOR

Old Bangkok
MICHAEL SMITHIES

Old Hanoi
MARK SIDEL

Old Kuala Lumpur
J. M. GULLICK

Old Kuching
ALICE YEN HO

Old Luang Prabang
BETTY GOSLING

Old Malacca
SARNIA HAYES HOYT

Old Manila
RAMÓN MA. ZARAGOZA

Old Penang
SARNIA HAYES HOYT

Old Singapore
MAYA JAYAPAL

Rice in South-East Asia: Cultures and Landscapes
JACQUELINE M. PIPER

Sarawak Crafts
HEIDI MUNAN

The Site of Angkor
JACQUES DUMARÇAY

Songbirds in Singapore
LESLEY LAYTON

Traditional Festivals in Thailand
RUTH GERSON

Votive Tablets in Thailand: Origin, Styles, and Uses
ML PATTARATORN CHIRAPRAVATI

Great Argus p. 8.

The Birds of Peninsular Malaysia

DEREK HOLMES

Illustrated by
KAREN PHILLIPPS

KUALA LUMPUR
OXFORD UNIVERSITY PRESS
OXFORD SINGAPORE NEW YORK
1998

Oxford University Press

Oxford New York
Athens Auckland Bangkok Bombay
Calcutta Cape Town Dar es Salaam Delhi
Florence Hong Kong Istanbul Karachi
Madras Madrid Melbourne Mexico City
Nairobi Paris Shah Alam Singapore
Taipei Tokyo Toronto

and associated companies in
Berlin Ibadan

Oxford is a trade mark of Oxford University Press

Published in the United States
by Oxford University Press, New York

First published 1998

British Library Cataloguing in Publication Data
Data available

Library of Congress Cataloging-in-Publication Data
Holmes, Derek (Derek A.)
The birds of Peninsular Malaysia/Derek Holmes; illustrated by Karen Phillipps.
p. cm.—(Images of Asia)
Includes index.
ISBN 983 56 0029 5
1. Birds—Malaysia—Malaya. I. Phillipps, Karen. II. Title. III. Series.
QL691.M4H65 1998
598'.09595' 1—dc21
97–42093
CIP

Typeset by Indah Photosetting Centre Sdn. Bhd., Malaysia
Printed by Kyodo Printing Co. (S) Pte. Ltd., Singapore
Published by Penerbit Fajar Bakti Sdn. Bhd. (008974-T),
under licence from Oxford University Press,
4 Jalan Pemaju U1/15, Seksyen U1, 40150 Shah Alam,
Selangor Darul Ehsan, Malaysia

Acknowledgements

THE author would like to acknowledge his appreciation to Ian Teesdale, who taught him so much about the birds of Malaysia during Sunday morning excursions in Johore in 1969–70. Secondly he wishes to thank David Wells who has readily given his time in correspondence on bird identification problems, and who willingly agreed to read through and comment on an early draft of the text. Lim Kim Seng and Sutari bin Supari have also both kindly consented to read and check the text and made valuable contributions, while Geoffrey Davison assisted in obtaining the Malay bird names and in confirming the current status of some species.

The illustrator would like to acknowledge the help of Dr C. M. Francis for help in the field and Dr K. C. Searle for checking the details and help with the references.

Jakarta
July 1997

DEREK HOLMES
KAREN PHILLIPPS

Contents

Introduction

My first experience of South-East Asian birds was in Brunei in 1968. Dropped into the jungle (literally, by helicopter) as a member of a team carrying out a reconnaissance soil survey of the entire state of Brunei, I was armed only with a copy of Smythies' *Birds of Borneo*, a battery-operated tape recorder, and a party of four Ibans. Each of them proved extremely useful in guiding me through the initial intricacies of identifying the birds of a forest avifauna that was almost completely alien to me. Otherwise, I was on my own, without contact at that time with the Malaysian ornithologists who might have guided me to many forest birds that I must surely have overlooked. I needed a book just such as this to introduce me to the unfamiliar families. For a young bird-watcher, it was an exciting time. Highlights of the memories were the Indian Cuckoo serenading our camp throughout the night in the rainy season with the opening bar of Beethoven's Fifth Symphony, the occasion when a pair of Black-and-Yellow Broadbills responded to a recorded play-back and sang furiously at me at arm's length; the first hearing of the astonishing call of the Helmeted Hornbill; and the final night in the forest when our Ibans captured a Sunda Ground-cuckoo for the pot, a record of this elusive bird that proved to be the first for Brunei. Perhaps the most lasting memory, however, so redolent of the lowland forests of the Malaysian region, is the incessant calls of the Blue-eared Barbet, at a time when these were sufficiently unfamiliar still to register on the ears of the newcomer. Invariably they were the last sound to be heard before the helicopter engines started up at the breezy coastal air base, and the first to be heard after the sound of the engines had died away, after leaving us at some airless clearing deep in the forest.

I returned to the region late in 1969, this time to the state of Johore in West Malaysia. For the development of my expertise with Malaysian birds, I shall be always grateful to Ian Teesdale, who guided me particularly to the tricky issues of recognizing secretive forest birds by their calls, such as the babblers, an art that was then still comparatively in its infancy. I also learnt from David Wells, though I was not to meet this former student-colleague until some four years later, at Satun and Songkhla Lake in southern Thailand.

Since then, I have studied birds mainly in Indonesia, especially in Sumatra and Kalimantan, where I further came to know and love the birds of the 'Sundanese' region (the subcontinental area that extends from the Thai Peninsula in the north, south to Java, and which has a broadly similar avifauna).

Specialist bird books and literature have always been available for the experienced ornithologist, but these are not readily assimilated by the beginner, or by those with a lively general interest in their natural surroundings. Throughout the years, I have felt the need for more popular bird books, which will help the beginner, as well as the more serious bird-watcher new to the region, around the intricacies of the bird families.

This book is the sixth in a series of popular bird books published by Oxford University Press in their *Images of Asia* series. It is designed to provide an introduction to most of the bird families that occur in West Malaysia. The selection of species within each family is never easy in a small book such as this, and emphasis is given particularly to describing the more colourful or distinctive birds that are peculiar to Malaysia. There are 146 species illustrated in colour, and a further 35 are illustrated by line drawings in the text; the systematic section refers to a total of 300 species.

East Malaysia

The birds described relate specifically to those of the Malay Peninsula, the region that extends from about Ranong in Thailand, south to Singapore. However, with certain precautions, this book can equally be used for the eastern states of Malaysia and for Brunei. An annotation (W) against a bird's name in the text indicates the few species that either do not occur in East Malaysia, or where present, are either very rare or local in distribution. Appendix 1 on p. 82 lists the species that are endemic to East Malaysia and the island of Borneo.

The English and scientific names used in this book mainly follow the nomenclature used in regional guides to South-East Asian birds. Malay names are given in Appendix 2. The subject of local names is complex, and further changes may occur. As a general rule, Malay names should be used with care. Often these names are recognized only in a general sense, for families or groups of birds. For example, *kelicap* is a commonly used word that can be applied to any small bird, while *pipit* is often not confined simply to the sparrows and munias. *Layang-layang* may be used for both swifts and swallows, which illustrates the concept that local names are based on superficial likenesses, and not upon genetic relationships.

Note

Abbreviations used on Plates:

♂	Male
♀	Female
Juv.	Juvenile

Sea Birds

THE ordinary land-based bird-watcher in Malaysia sees very few sea birds, apart from small flocks of terns (see p. 13), and occasional frigate birds. The sea bird colonies in our adjacent seas have been severely depleted by the activities of visiting fishermen.

BROWN BOOBY

Sula leucogaster (75 cm)

Brown Booby

One sea bird that can be claimed as a Malaysian resident is the Brown Booby. Boobies are related to the gannets of temperate regions: large, cigar-shaped birds with long, narrow wings and long, heavy bills. They feed by plunging into the sea for fish, generally from a low angle (unlike gannets which plunge vertically), and they fly low over the surface with steady wing beats alternating with short glides. Mainly brown, the adult Brown Booby has a white belly and underwing. Pulau Perak formerly had a colony of 4,500–5,000 pairs, but the depredations of fishermen have reduced this to around 100 pairs at the last count.

There has been no recent survey of this island but the colony's extinction seems imminent. There are small colonies on islands in the South China Sea such as Pulau Layang-Layang.

Herons, Egrets, and Bitterns

THESE are the tall, long-necked and long-legged wading birds of the coastal marshes and wet rice-fields, which have long, dagger-like bills used for stabbing at fish and crustaceans. In flight, they hold their necks in an 'S' shape, and fly with rather slow, steady beats on rounded wings.

CATTLE EGRET
Bubulcus ibis (50 cm)

The egrets are the graceful white herons of coastal marshes and mud-banks, and are mainly migrants to our region. The Cattle Egret is the smallest, and is partial to feeding near grazing cattle. It has a yellow bill and acquires some buff plumes before departing in spring. A related bird in the rice-fields is the CHINESE POND-HERON *Ardeola bacchus* (45 cm). This is mostly seen in Malaysia in its winter plumage, which is a rather cryptic pale brown, but with startling white wings when it takes flight.

Cattle Egret

Only two egrets breed in the peninsula, the stately, white GREAT EGRET *Casmerodius albus* (90 cm) at isolated colonies on the west coast, and the REEF EGRET *Egretta sacra* (58 cm). The Reef Egret occurs on muddy shorelines and coral reefs, and has two quite different colour phases; one is slaty grey and readily recognized, while the other is pure white, and requires care to distinguish from the other egrets.

Reef Egret

The GREY HERON (W) *Ardea cinerea* (100 cm) is light grey, with a white head and neck. Long black plumes arise from behind the eyes. It is a resident of the muddy west coast. The smaller PURPLE HERON *A. purpurea* (97 cm) is grey and purple-brown in colour, and can be found widely in coastal rice-fields.

The LITTLE (or Striated) HERON *Butorides striatus* (46 cm) is a small, rather hunched heron that lives along the densely vegetated sides of creeks, both in the mangroves and along freshwater rivers in wooded regions. It ranges widely over the mud-flats exposed at low tide, but is a wary bird, gently flying off low over the surface before an intruder can get near. It is quite a common resident, with numbers augmented in winter.

Little Heron

The bitterns are more secretive birds of freshwater reed-beds and swamp scrub, typically seen when they make short flights on rounded wings, with hunched necks. The commonest species is the

CINNAMON BITTERN *Ixobrychus cinnamomeus* (40 cm) which is rather uniform chestnut in colour. It is both a resident and a winter migrant, and can be found almost anywhere in swampy thickets in open country.

Storks

STORKS are tall, heavy, gaunt birds with long legs and necks and massive, dagger-like beaks. They are experts at soaring in thermals, sometimes to a considerable height. All storks in Malaysia are endangered, such large and conspicuous birds being unable to compete with Man's rapid development.

The LESSER ADJUTANT *Leptoptilos javanicus* (115 cm) is a large and evil-looking bird that can sometimes be seen stalking grotesquely in open ground along estuaries, coastal rice-fields, and river valleys, usually singly or in twos or threes. Few of its breeding sites are known. It is mainly black above and white beneath, and it has a massive bill. Populations are declining everywhere, and the last estimate was that there are no more than 200 birds along the length of the west coast, with very few on the east coast.

The MILKY STORK (W) *Mycteria cinerea* (100 cm) is white, except for a dark breast band and black tail and flight feathers. The bill is very slightly down-curved. This stork feeds largely on mud-skippers in the very soft mud of the inter-tidal zone, walking very deliberately, a few paces at a time, probing its bill deep into the mud. The species is endangered and very few now breed in the Malay Peninsula. About 100 birds still occur along the west coast, mostly in Perak. However, there remains a substantial population along the east coast of Sumatra.

Birds of Prey

THE birds of prey, or raptors, form quite a large and diverse group of predatory birds. All have hooked bills and strong talons, but the methods of hunting are varied, including soaring, hovering, diving, quartering the ground, and swift flight through the canopy. A notable feature is the mass migrations of certain species, soaring high on the thermals, southwards down the peninsula in October and November, many debouching across the Straits of Malacca at Cape Rachado in Negeri Sembilan. The flight back north is in March and April.

BRAHMINY KITE
Haliastur indus (45 cm) Plate 1a

This chestnut and white bird is a familiar resident of rivers, coasts, and harbours, and is generally the identity of the 'sea eagle' reported to ornithologists by their friends. It is not an eagle, although it has similarly broad wings and tail, but unlike most kites, the tail is not forked. The adult shown in Plate 1a is distinctive, but the young bird is dull brown and spotted, and so may be confusing. The call is shrill and quavering.

The WHITE-BELLIED SEA-EAGLE *Haliaeetus leucogaster* (70 cm) is a much larger bird of the coasts. Its short, wedge-shaped tail is distinctive, and the under-parts in the adult are mainly white.

White-bellied Sea-eagle

BLACK-SHOULDERED KITE (W)
Elanus caeruleus (33 cm) Plate 1b

The Black-shouldered Kite is a graceful raptor, usually seen hovering over open country. It is pale grey, whiter below, with a distinctive black patch on the bend of the pointed wing. This is the only resident raptor that hunts by hovering.

CRESTED SERPENT-EAGLE
Spilornis cheela (60 cm) Plate 1c

The novice may have difficulty in identifying the several eagle species that occur, but the Crested Serpent-eagle is the most widely distributed, and readily identified in overhead flight by the single broad white band in the tail and another close to the hind margin of the underwing. The call is a shrill di- or tri-syllabic note which often draws attention to a soaring bird or pair. When the bird is perched, an observer should look for the slight drooping crest, although this is not a diagnostic feature.

Among the larger eagles, the two most common species are the BLACK EAGLE *Ictinaetus malayensis* (70 cm) and the CHANGEABLE HAWK-EAGLE *Spizaetus cirrhatus* (70 cm). The Black Eagle appears all black, with yellow feet, and occurs mostly in the mountainous areas. The Changeable Hawk-eagle is usually seen soaring over lowland forests, with very broad wings and a long, rounded tail; the plumage is variable, but most birds are brown above and paler below, and there are fine dark bars on the underwings and tail.

JAPANESE SPARROWHAWK
Accipiter gularis (27 cm) Plate 1d

The sparrowhawks have short, broad wings and rather long tails, and hunt by rapid flight at or below tree-top level, preying mainly on small birds. Even larger pigeons fly off in alarm when one passes overhead. The Japanese Sparrowhawk is a common migrant during the northern winter, and is the species most often encountered, but the different species are not easy to identify. The upper-parts are grey in the male and brown in the female, and the under-parts are barred brown and white.

This is one of the common constituents of the annual migratory stream mentioned above. Another abundant member is the ORIENTAL HONEY-BUZZARD *Pernis ptilorhynchus* (50 cm), a large raptor with a noticeably small head. The under-tail pattern aids identification: white with two dark bands near the base and another at the tip.

BLACK-THIGHED FALCONET
Microhierax fringillarius (15 cm)

Casual bird-watchers will often overlook the falconets, not realizing that such small birds are falcons, but their ferocity and hooked bills will soon reveal their affinity. The falconet occurs on forest edges and in lightly wooded country, where it perches openly in low trees, making short, direct flights. The plumage, mainly black above and white with some rufous below, is attractive, and in their unobtrusive way, these birds are quite distinctive.

Black-thighed Falconet

Ducks

LESSER WHISTLING-DUCK
Dendrocygna javanica (40 cm) Plate 2a

The Malay Peninsula has very few species of duck, and this is the only species that is confirmed as a breeding resident locally. It is so named from its mellow, whistling calls as small flocks flight between feeding grounds in swamps, wet rice-fields, lakes, and mining pools. It is an all brown duck, more chestnut on the wings, and with a dark cap. In flight, the head is generally held slightly lower than the body.

Game Birds

THE game birds include the partridges and pheasants, but visitors from more northern countries should realize that our forest-dwelling species are extremely secretive and difficult to study, and only the more vocal forms readily reveal their presence.

GREAT ARGUS

Argusianus argus (75–200 cm) Frontispiece

Although an observer may never see one, the Argus is familiar because of its loud calls. These can be heard commonly in lowland and hill forest, and typically consist of a loud, far carrying 'ku-wao'. Another call is a long series of up to thirty mono- or slightly di-syllabic notes in a series on one pitch, becoming slower and rising slightly towards the end. Surveyors in the forest will sometimes come across the male's dancing grounds, circular clearings of 3 metres or more in diameter, from which every twig and leaf have been removed; he uses this for his dancing displays to his mate. The length given above includes the tail of the male bird, which can reach up to 150 centimetres. The secondary feathers of the male's wings are likewise enormously lengthened and decorative, with beautiful eye-like markings and patterns of rufous, buff, and black.

The CRESTED ARGUS (W) *Rheinardia ocellata* is a rare bird in the hill forests up to about 1000 metres. Instead of the elongated secondaries, the male has an enormous broad, long tail. It is named from its long, white crest.

CRESTLESS FIREBACK

Lophura erythrophthalma (50 cm) Plate 2b

The other pheasants of the forest are much less vocal and therefore less conspicuous. Nobody really knows how common they are but those confined to lowland forest must surely be treated as threatened. Both this species, which is shown in Plate 2b, and the CRESTED FIREBACK *L. ignita* (70 cm) are lowland pheasants. The male Crestless Fireback is bluish-black, but with red facial skin and back and orange-coloured tail feathers, while the female is all black except for the facial skin. The male Crested Fireback differs in having an upright tufted crest, blue facial skin, and a whiter tail, while the female is chestnut with white scalloping below.

CRESTED WOOD-PARTRIDGE

Rollulus rouloul (25 cm) Plate 2d

Among the smaller partridges, which are occasionally seen as small parties of black forms running or whirring away through the undergrowth, the Crested Wood-partridge (or Roulroul) may be quite common. Its melancholy slurred whistles sometimes reveal its presence. Plate 2d shows the large crest and glossy colour of the male, and the greener plumage of the female with brown wings.

The LONG-BILLED PARTRIDGE *Rhizothera longirostris* (30 cm) can be a noisy bird in the foothills and at some of the hill stations. It has a shrill double whistle, and duetting pairs will produce a rising whistle of four to six syllables.

Often heard in forest edge habitats is the RED JUNGLEFOWL (W) *Gallus gallus*, which crows very like the domestic cockerel though with a rather thinner and more abrupt quality.

BLUE-BREASTED QUAIL
Coturnix chinensis (15 cm) Plate 2c

A much smaller game bird, the Blue-breasted Quail is common in open grasslands, dryland crops, and ground scrub. The male is distinctively marked but it is very difficult to obtain a good view. Generally it runs to hide, or leaps up on approach and is seen only as a whirring brown ball diving into cover. The female is dull brown.

Crakes and Rails

MOST crakes and rails are secretive birds of the marshes, difficult to observe as they quickly run for cover in the reed-beds or wet thickets when disturbed. They fly weakly, with dangling legs. The short tails are frequently flicked up, usually revealing a striking white, rufous or barred patch under the tail.

WHITE-BREASTED WATERHEN
Amaurornis phoenicurus (33 cm) Plate 3b

This is quite a conspicuous bird that is often seen in a variety of damp locations such as roadside ditches, thickets, and rice-fields. The plumage is distinctive and it frequently cocks its tail as it walks, revealing the chestnut under-tail coverts. The call is a prolonged cacophony of grunts and chuckles that develop into a series of 'kru-wak kru-wak' calls, sometimes for minutes on end.

The MOORHEN *Gallinula chloropus* (30 cm) is quite a common rail usually seen swimming at the edge of reed-fringed ponds and mining pools, or the wetter rice-fields. It is all black except for white patches under the tail and some white along the flanks. It has a bright red frontal shield and a short bill. It commonly reveals its presence by its short liquid 'prk' call.

WHITE-BROWED CRAKE
Poliolimnas cinerea (20 cm) Plate 3a

The White-browed Crake is the most readily seen of the smaller rails and crakes, as it often comes into the open areas adjacent to the reeds and rushes of the small marshes that it frequents. Brown above, grading to whitish on the under-parts and buff under the tail, its diagnostic feature is the grey head and neck with two prominent white streaks on the face, above and below the eye. The legs are greenish. It is quite vocal, with shrill, piping 'weeu, weeu, weeu' calls, especially early and late in the day.

Jacanas

PHEASANT-TAILED JACANA (W)
Hydrophasianus chirurgus (30 cm) Plate 2e

Although not common, jacanas are very striking, with their long legs, toes, and claws, specialized for walking on floating vegetation in swamps, earning them the name of 'lily-trotters'. They can swim and even dive if required. The flight is weak, and like rails, the legs trail behind, and the wings are short and rounded. The Pheasant-tailed Jacana is a winter visitor from the north, at which season it is mostly brown above and white below, with a dark breast band and yellow on the hind-neck. In flight, the wings are wholly white. Before they depart north in about May, the males begin to exhibit the beautiful plumage shown in Plate 2e, with a tail of 25 centimetres. The largest gathering so far reported in Malaysia was a party of over forty seen on migration by the author on the Sedili Kecil River in eastern Johore.

A major prize for any bird-watcher is a glimpse of the rare MASKED FINFOOT (W) *Heliopais chirurgus* (53 cm). The author has seen it just once, on Ayer Keroh Lake at Malacca. It is a swimming bird, confined to overgrown water-side habitats, whether these be the sides of swift, clear rivers, backwaters, or mangroves. Its rather elongated body sits low in the water, with the neck bobbing as it swims. Its thick, long bill recalls a banana stuck on the front of the head. All records date from December to June, but it has not yet been positively confirmed that all are of migrants.

Shore Birds

THIS group includes all the brown wading birds that are found around the estuaries and tidal mud-flats of the coast and on muddy patches inland, mostly during the northern winter and on migration. They range in size from the stints (15 cm) to the tall greenshanks (35 cm) and long-billed curlews (58 cm), but the identification of the smaller species in particular can be difficult, and is beyond the scope of this book.

COMMON SANDPIPER

Actitis hypoleucos (20 cm) Plate 3c

Perhaps the most familiar of these waders is the Common Sandpiper, which is a common migrant to water-side habitats everywhere, whether coasts or harbours, mangroves, inland pools, or rivers in the hills. It is not averse to perching on rocks or concrete piers, or resting on floating logs. It constantly bobs its body up and down when feeding. In its short, jerky flights, usually accompanied by a shrill 'twee-wee-wee', it alternates rapid wing-beats with short glides on down-curved wings, showing its prominent white wing bar and white outer tail feathers. It is seen mostly from September to April.

Quite common in small parties in rice-fields is the WOOD SANDPIPER *Tringa glareola* (23 cm). This bird is completely dark above, except for the prominent white upper tail coverts seen in flight. The under-parts are whitish, but the wings are all dark without a white bar. Its plaintive 'tiss-iss-iss' call often rings across the rice-fields during the northern winter.

GOLDEN PLOVER

Pluvialis fulva (25 cm) Plate 3d

Another migrant wader that is easily identified is the Golden Plover, a bird that especially favours short grass. From October to April, residents near golf-courses, football pitches, and airfields will often hear their whistled 'tu-ee' calls as they fly in at night to roost. The plumage is brown and slightly mottled, paler below and with unpatterned wings. The short bill is dark. Prior to departure north to the breeding grounds, some birds begin to acquire summer plumage, which has golden mottles and black under-parts.

WHIMBREL

Numenius phaeopus (43 cm)

Among the many wader species that visit the coastal mud-flats, the large Whimbrel is identified by its very long, down-curved bill. This bird particularly favours the mud-flats adjacent to mangrove swamps, and it will on occasion wander deep into the mangroves. Brown above and paler below, it has a white triangle extending up the back, which is prominent in flight. The call is a rapid, piping 'pi-pi-pi-pi'.

Whimbrel

ORIENTAL PRATINCOLE
Glareola maldivarum (25 cm) Plate 3e

Although related to the waders, this attractive bird is an aerial insect feeder, and behaves more like a swallow, with its slightly curved wings and forked tail. It is mainly a passage migrant, passing south in quite large numbers in September to November and returning in February to March, though some may breed in the northernmost states. It is particularly attracted to the short grass of airfields. At close range, the dark bordered buff throat, chestnut linings of the underwing, and white lower back are distinctive.

Terns

TERNS are graceful birds of the open seas, coasts, and more rarely, inland waters. They have narrow wings, long bills, and rather long, forked tails. They feed by diving into the water, sometimes hovering first (*Sterna* sp.), or by quartering over marshy areas, plucking food from the surface but not diving (the marsh terns, *Chlidonias* sp.).

BLACK-NAPED TERN
Sterna sumatrana (30 cm)

The Black-naped Tern is distributed all round the coasts, more especially the rocky shores. It does not congregate in flocks like other terns, nor does it occur inland. It breeds on off-shore rocky islands, and like all terns, will often rest on fishing platforms. It is whiter than most terns, with only a narrow black collar around the back of the nape. This tern has a black bill and long tail streamers which give the tail a deeply forked appearance. The only other tern that breeds close to the mainland is the graceful LITTLE TERN *S. albifrons* (25 cm), which is found particularly along the estuaries on the east coast.

Black-naped Tern

GREAT CRESTED TERN

Sterna bergii (46 cm)

This is perhaps the most generally distributed tern in our waters. It is a large tern with black crown, white forehead, and a lemon-yellow bill. A short crest is visible when perched. The similar LESSER CRESTED TERN *S. bengalensis* (38 cm) differs mainly in having an orange-yellow bill.

Great Created Tern

WHITE-WINGED TERN

Chlidonias leucopterus (25 cm)

The marsh terns favour freshwater ponds or swampy rice-fields, brackish-water marshes, and shallow muddy tidal waters. They feed by quartering restlessly back and forth at low height and plucking insects from the surface. They are not quite as graceful and slender as sea-terns, and their tails are only slightly forked. In breeding plumage, the head and body are black, and the rump, tail, and upperwings are white, but most birds in Malaysian waters are in their winter plumage, mainly white and

White-winged Tern

grey with a black patch on the rear of the crown and ear coverts. The experts should be aware that the WHISKERED TERN *C. hybridus* is also a winter visitor. The two species are very difficult to distinguish until they begin to develop their breeding plumage prior to departure back north.

Pigeons

THERE are some twenty species of pigeons and doves in our region, but they fall into a number of distinct genera, and identification of most species is not difficult if good views are obtained. Some are quite colourful and their bubbling or cooing calls are attractive. Many are forest birds, but being strong flyers, they range quite widely over a variety of habitats.

LITTLE GREEN PIGEON

Treron olax (20 cm) Plate 4a

The green pigeons are plump fruit eaters of the forest. The females are mostly green, but the males are more diversely coloured, with maroon mantle and wings in some species. The best place to observe them is at fruiting fig trees, when they will congregate greedily together with barbets and hornbills. Plate 4a shows that the male Little Green Pigeon is quite a colourful bird. It frequents the canopy of lowland forest, where the expert can identify it by its call. Most green pigeons have a short chuckling, cooing note or gurgle, but in this species the call is preceded by a nasal winding-up whine.

Another common lowland species is the THICK-BILLED PIGEON *T. curvirostra* (27 cm), in which the male also has the maroon mantle, but is otherwise mainly green, with a greyish cap.

PINK-NECKED PIGEON

Treron vernans (27 cm) Plate 4b

This is the green pigeon that is most often found in secondary growth of more open country, and is one species in which the male has a green rather than a maroon mantle. The fore-parts resemble those of the Little Green Pigeon, but it has a broad, pale pink collar and breast, above an orange lower breast band.

A rather striking species found in the poorer forests on lowland acid soils, particularly in the south and east, is the CINNAMON-HEADED PIGEON *T. fulvicollis* (25 cm), in which the male is mainly reddish, ranging from chestnut on the fore-parts to maroon on the mantle. By contrast the female is almost entirely green.

MOUNTAIN IMPERIAL PIGEON

Ducula badia (47 cm) Plate 4d

The Imperial Pigeons are heavier birds of the forest canopy. This species is pinkish-grey, with a reddish-brown mantle, buff under the tail, and a pale terminal half of the tail. It is most often seen in flight, especially in the morning and evening, when its booming call, 'hoo-HOOM', echoes across the mountain valleys.

The GREEN IMPERIAL PIGEON *D. aenea* (43 cm) replaces it in the lowland forest, but is now a rare and very local bird. It has a green mantle and tail, and its deep, soft 'hroom' call, preceded by a quiet clicking note only audible at close quarters, is one of the characteristic sounds in those places where it occurs, such as the Endau–Rompin reserve.

LITTLE CUCKOO-DOVE

Macropygia ruficeps (30 cm) Plate 4c

The Cuckoo-doves are slender, rufous-brown doves with long tails. They have swift flight and frequently all that is seen of them is two or three birds darting above the canopy on the sides of a valley in hilly country. The plumage of this species is mottled rufous and faintly barred. The call is a rapidly repeated 'croo-wuk, croo-wuk', though at a distance it sounds as 'wuk, wuk, wuk---'.

SPOTTED DOVE

Streptopelia chinensis (30 cm) Plate 4f

This turtle-dove is a common inhabitant of open country and even the wooded suburban areas of towns. It is most often seen flying up from the roadsides and bare ground near bushes, when the black-and-white spotted half-collar and broad white tips to the outer tail feathers are clearly seen. The call consists of three or four throaty cooing notes.

PEACEFUL DOVE

Geopelia striata (21 cm) Plate 4e

This small, slender, long-tailed dove is noted for its musical call, a series of rapid trilled cooing notes. It prefers more open wooded areas, such as the scrub growth at the back of sandy beaches, where it is predominantly a ground bird, using low and partially hidden perches. It has irregular black barring on the upper-parts, neck, and flanks, from which it is also known as 'Zebra Dove'.

JAMBU FRUIT-DOVE

Ptilinopus jambu (27 cm) Plate 5a

The Jambu Fruit-dove is an unobtrusive but colourful inhabitant of the middle storey of lowland and hill forest. It is usually chanced upon by accident, when a bird makes a brief flight on one's approach, but then perches to give a good view. The female is less colourful, with a green breast and white belly. The white eye-ring is distinctive in both sexes. Generally a silent bird, it will occasionally give a low 'coo', audible only at rather close range.

GREEN-WINGED PIGEON

Chalcophaps indica (25 cm) Plate 5b

The Green-winged Pigeon (also known as Emerald Dove) is a common inhabitant of the understorey, feeding on the ground. It occurs in forest, secondary woodland, and old rubber plantations, and is often flushed from the side of logging roads. The usual view is of a metallic green bird in swift, low flight away from the observer, when two strong white bars are prominently seen on the lower back. The female is somewhat duller. It is rarely possible to get a good view of the bird on the ground, but hardly a day passes in the forest without its call being heard, a soft call that sounds like 'wup, wup, wup' uttered from the understorey. At very close range, it is heard to consist of a double note, 'cu-wup'.

Parrots

PARROTS are characterized by their heavy hooked bills used for tearing open fruits and nuts, and also as a grip for climbing along branches. Malaysia has only three species.

LONG-TAILED PARAKEET

Psittacula longicauda (40 cm, with tail) Plate 5d

Although the male of this parrot is quite colourful, generally it is only seen as a long-tailed green bird in raucous parties in the canopy. It is the commonest member of the family throughout the region, but numbers appear to fluctuate according to poorly understood seasonal movements. Roosting flights over the forest in the evenings may indicate the presence of some quite large roosts, especially near the coasts. Both this and the next species sometimes frequent oil palm estates.

BLUE-RUMPED PARROT

Psittinus cyanurus (19 cm) Plate 5c

This short-tailed parrot is not as common as the parakeets, though a few will be heard on most days spent in the forest. The call is a shrill, sharp whistled note or series of notes, surprisingly musical for a parrot, usually heard from birds in flight at canopy level. The female is a duller bird, with a brownish head, and it lacks the red bill.

BLUE-CROWNED HANGING PARROT

Loriculus galgulus (14 cm) Plate 5e

The hanging parrots are charming birds, stumpy, seemingly almost neckless little bundles of bright green that always rest hanging upside-down. They clamber about the outer branches with great dexterity, but are most often seen in swift flight, giving high, thin 'zri-ie' calls. The female lacks the bright patches of colour of the male, except for the scarlet rump. It is common in the lowlands and hills, mainly in the forest.

Cuckoos

THE cuckoos form a large group—our region has fourteen species of the true cuckoos, the birds which are well known as brood parasites, laying their eggs in the nests of other birds and leaving the raising of their young in the care of the foster parents. In addition, there are six species of malkoha and three coucals. Unobtrusive when not singing, cuckoos are most easily identified from their songs.

PLAINTIVE CUCKOO

Cacomantis merulinus (22 cm) Plate 6a

The Plaintive Cuckoo is a familiar bird, thanks to its voice. Found in lightly wooded open country, it favours low perches and even telephone wires and fences, so is not too difficult to observe. It can also be heard in towns and suburbs. The rather long, graduated tail (rounded when fanned) is typical of the cuckoos, as are the slightly pointed wings. Quite small for a cuckoo, the adult plumage in Plate 6a would appear to be distinctive, but the Brush Cuckoo (see below) is very similar, and the best identification is from their high pitched, whistled songs. There are two quite separate songs which can be interchanged at will. In the Plaintive Cuckoo, there is a diagnostic cadence song, beginning with four slow notes then breaking into eight rapid notes down the scale. The second song consists of the phrase 'tay-ta-wi' repeated about four times up the scale. The usual foster parents are believed to be the small ioras and warblers.

The notes that follow provide a simple guide to the other cuckoos most often encountered, particularly their calls.

BRUSH CUCKOO
Cacomantis sepulcralis (24 cm)

The rising song of the Brush (or Rusty-breasted) Cuckoo is the same as that of the Plaintive Cuckoo, but the cadence is quite different, consisting of a long series of at least ten plain whistles, 'heet, heet, heet, ...' at the same speed, falling very slightly in pitch. A darker colour than the Plaintive Cuckoo, it is more of a woodland or forest bird, rarely coming so far out into open country.

BANDED BAY CUCKOO
Cacomantis sonneratii (23 cm)

The Banded Bay Cuckoo is a bird of the forest canopy, and sometimes plantations, where it is difficult to see. Its song consists always of four thin, shrill whistles, falling slightly but evenly in pitch in a minor key. This song should be compared with that of the Indian Cuckoo.

INDIAN CUCKOO
Cuculus micropterus (30 cm) Plate 6b

The Indian Cuckoo especially favours the canopy of forest margins in wet places such as river valleys and freshwater swamps. Like the Banded Bay Cuckoo, it also has a song which invariably consists of four notes, but in the Indian Cuckoo the notes are much more emphatic, three notes almost on one pitch and the fourth lower, recalling the opening bar of Beethoven's Fifth Symphony. This call is a frequent accompaniment to evenings in a forest camp in the lowlands, sometimes heard throughout the day and night.

LARGE HAWK-CUCKOO
Cuculus sparverioides (33 cm)

This bird of forest margins in the mountains above 1000 metres is remarkable for its song, from which it has earned the name 'Brain-fever bird' (though that name has been applied to several cuckoos). The typical song is a double note, 'pi-pee', that is repeated faster and faster up the scale until it reaches a frantic crescendo. A similar call consists of a single note that is repeated at constant speed rapidly up to a crescendo and then fading off. The two songs can be interchanged at will. The effect is stunning, though two birds singing in competition throughout a moonlit night can be very wearying.

VIOLET CUCKOO

Chrysococcyx xanthorhynchus (16 cm)

This bird of lowland forests is most often seen as a dark cuckoo in song-flight above the forest canopy or across logging roads; the flight is dipping, the dips synchronized with a rather shrill 'kie-vik', one of the common sounds of the forest although easily overlooked by any-one not yet familiar with it.

DRONGO CUCKOO

Surniculus lugubris (23 cm)

One of the characteristic calls of lowland forests and woodlands is the human-like whistle of five to seven notes rising the scale, given by the Drongo Cuckoo. It is a black, drongo-like bird, with a very slight fork in its rather long tail, which is barred with white.

COMMON KOEL (W)

Eudynamys scolopacea (43 cm)

A large and heavy cuckoo, black in the male and brown with paler markings in the female, this cuckoo is at once identified by the call, a loud 'ko-el' that rises in pitch and crescendo; it also makes excited bubbling calls. It parasitizes the nests of crows, and was formerly rare, and mainly a migrant, but the resident population appears to be growing rapidly wherever House Crows (see page 80) have become common, such as in Penang and the Kuala Lumpur region.

RAFFLES' MALKOHA

Rhinortha chlorophaea (33 cm) Plate 6c

The malkohas are not parasitic and appear to have little in common with the true cuckoos. All are characterized by their very long tails. They are rather sluggish birds of the middle storey that tend to creep around the foliage rather like squirrels, occasionally making short, rather laboured flights to another tree, on broad, rounded wings.

With six species, it is difficult to select any one that is typical. The Raffles' Malkoha is one of the commoner species in the lowlands, in both primary and secondary forest. It is the smallest, and the only species in which both the bill and the eye-shield are green. In contrast to other species, the sexes differ strikingly, the male being mainly chestnut, while the grey fore-parts of the female are very distinctive. Most malkohas are generally silent, but this species habitually gives a series of about five mewing notes down the scale, uttered as if with great effort.

Two other malkohas are also mainly chestnut. The CHESTNUT-BREASTED MALKOHA *Rhamphococcyx curvirostris* (46 cm) has a green bill and red eye-shield, chestnut breast and belly, a greenish mantle, and no white in the tail. It is common in open scrub. The RED-BILLED MALKOHA *Zanclostomus javanicus* (46 cm) is the only species with a red bill, and it has no obvious eye-shield.

CHESTNUT-BELLIED MALKOHA
Rhopodytes sumatranus (40 cm) Plate 6d

There are three species which are mainly grey, with green bill, red eye-shield, and white tips to the tail feathers. The grey is quite dark, with a greenish gloss. This species has chestnut under-tail coverts, and although not easy to see, these distinguish it from the very similar BLACK-BELLIED MALKOHA *R. diardi* (37 cm). The GREEN-BILLED MALKOHA (W) *R. tristis* (56 cm) is a much larger bird of hill forest, and also more open lowlands in the north of the peninsula.

GREATER COUCAL
Centropus sinensis (53 cm) Plate 6e

The coucals or crow-pheasants are rather heavy, clumsy birds with medium-length tails. The Greater Coucal is common in dense thickets, whether in primary or secondary forest or damp places such as the edge of rice-fields. It is readily identified, with its thick bill, black and chestnut plumage, and graduated black tail. It is a skulking bird, often feeding on the ground, and its local name of 'butbut' is derived from the call, a number of hollow 'hoop' notes. The full song consists of a series of about thirty such notes, beginning loud and quite fast, but soon tailing off into a long series of even notes.

In thin secondary scrub and bushes in dry grassland, one finds the LESSER COUCAL *C. bengalensis* (38 cm), a rather more scruffy bird with dark brown or black body, much spotted and streaked with white. The usual call is three or four 'hoop' notes that break into a staccato call, quite distinct from that of the Greater Coucal.

Owls

THE mysterious owls and other night birds are intriguing, but difficult to study. Owls range in size from the small scops owls to the large eagle-owls. All of them are primarily arboreal and nocturnal, but an observer will sometimes chance upon an owl during the daytime, usually sitting silently, suffering with resignation the mobbing of small birds that have revealed its presence. One of the disconcerting features of owls is their ability to rotate their heads, always to face the intruder, with their big round eyes set in a flat facial disc.

COLLARED SCOPS OWL
Otus bakkamoena (22 cm) Plate 7a

The Collared Scops Owl occurs commonly in open country, and its single, slightly querying 'wok' calls, uttered at irregular intervals, are often heard in wooded villages or the tree-lined suburbs of towns. A torch light will reveal the yellowish-brown eyes and short ear tufts.

COLLARED OWLET
Glaucidium brodiei (17 cm) Plate 7f

This tiny owl of the mountain forests is partially diurnal, and has a distinctive ventriloquial four-note call, the middle two notes short and run together. Another montane owl is the MOUNTAIN SCOPS OWL *Otus spilocephalus* (20 cm), which has a plaintive double whistle, 'plew, plew'.

BROWN HAWK-OWL
Ninox scutulata (30 cm) Plate 7d

Perhaps the most frequently heard owl in forest is the Brown Hawk-owl, so called because it is rather hawk-like in appearance, lacking the flat facial disc of most owls. It has a hawk's powerful hooked bill and glaring yellow eyes. It lives in all kinds of forest in the lowlands and lower hills, where its call can be heard most nights, a repeated double note, 'wu-wup, wu-wup, wu-wup', the second note higher and emphasized, uttered more rapidly than the irregular, nearly monosyllabic calls of the Collared Scops Owl.

BARN OWL (W)
Tyto alba (34 cm) Plate 7b

This is a ghostly coloured owl that hunts on completely silent wings over open country, breeding in big old trees and large or disused buildings. Formerly a very rare bird, deforestation and especially the development of rat-infested oil palm estates encouraged a population explosion of this species in the 1970s, and it is now widely distributed. Its eerie shrilling call enhances its ghostly character, but it is truly a farmer's friend.

BUFFY FISH-OWL
Ketupa ketupu (45 cm) Plate 7c

Among the larger owls, perhaps this is most often seen, as it favours the more open wooded localities near water, sometimes flying alongside rivers or swamps, and even occasionally rice-fields in wooded country. It feeds on fish, frogs, and other water creatures, and its legs are unfeathered, appropriate to a watery environment. It is a rich rufous colour with heavy dark vertical streaks. Prominent ear tufts (tufts of feathers) lie on the sides of the head.

In the forest, the rather similar BARRED EAGLE-OWL *Bubo sumatranus* (45 cm) has greyer plumage, prominent ear tufts, and fine horizontal bars on the breast. It is one of the authors of deep owl hoots in the forest.

Nightjars

NIGHTJARS are nocturnal aerial feeders, with long, pointed wings, rather long tails, and acrobatic, hawk-like flight. The rich, heavily mottled brown plumage is a good camouflage for birds resting on the ground or along branches, usually lengthwise, during the day. The delicate bill opens to reveal a wide gape for catching insects in flight.

MALAYSIAN EARED NIGHTJAR
Eurostopodus temminckii (27 cm) Plate 7g

This is the common forest nightjar of the lowlands and lower hills. It lives in the forest, and precisely at dusk and again at dawn, all the eared nightjars in the district will be in flight, above the canopy or around clearings, giving their unmistakable 'tok, tedau' calls. The dusk call comes just as the sunset shrilling of the cicadas dies down, but before full darkness sets in. Unlike other nightjars and owls, they are never heard through the night. Note the prominent ear tufts and pale buff collar in this species.

The range of this species overlaps along the northern border with that of the GREAT-EARED NIGHTJAR (W) *E. macrotis* (40 cm). The call of this much larger bird is a loud 'peeu-weeu', preceded by a softer 'pip'.

The open parklands, even the suburbs of towns, have the LARGE-TAILED NIGHTJAR *Caprimulgus macrurus* (30 cm) which has a monotonous, emphatic, and rather resonant 'tonk-tonk, tonk-tonk-tonk' call. This regular call outside one's bedroom can be irritating, especially when two birds are calling out of synchrony.

JAVAN FROGMOUTH
Batrachostomus javensis (24 cm) Plate 7e

The frogmouths are an interesting nocturnal family about which very little is known. Superficially similar to nightjars, they derive their name from the bill, which is of almost ridiculous width. They are a rich, mottled rufous in colour. Often silent birds, the calls are poorly known but can be a guide to identification. The Javan Frogmouth has a ventriloquial, mournful, soft, falling whistle, wavering slightly at the end, uttered at intervals through the night. Mainly a forest bird, the author has seen it also in secondary growth and old rubber trees.

Swifts

SWIFTS are aerial insect-feeders that spend the greater part of their lives in the air. So common at any one place by day, it is often a puzzle where they spend the night. Being masters of the air, true swifts have very weak legs, and never perch like swallows on wires, branches, or roof-tops, but cling to vertical surfaces. The largest swifts are capable of extremely rapid flight and one may be surprised by the 'whoosh' of a passing needletail. Mainly blackish birds, they have curved, sickle-shaped wings, and fly with alternating wing beats and tilting glides, never closing the wings like the superficially similar swallows.

The small swiftlets and the large needletails are notoriously difficult to identify. It is from some of the swiftlets that come the birds-nest soup (nests constructed of saliva), and the cave-dwelling members of the group can navigate by echo-location. A good time to watch swifts is when one chances upon them dipping down to the surface of rivers or ponds to drink.

HOUSE SWIFT (Little Swift)
Apus affinis (15 cm)

This is the common swift of towns everywhere, with sometimes hundreds nesting in the eaves of downtown shop-houses. It is the first bird to greet visitors arriving at the Kuala Lumpur airport. A blackish bird, it has slender, curved wings and a slightly forked tail. The rump and throat are white. It is a noisy bird, and in the evenings especially, the streets are full of whirling, twittering birds. This bird is continuing a population explosion that is probably a response

House Swift

to the substitution of man-made structures for its original rock-nesting sites.

The PALM SWIFT *Cypsiurus balasiensis* (13 cm) is a smaller, very slender swift, dark brown with very narrow, curved wings and rather long, forked tail, that is generally seen around palm trees in open country. It has a distinctive, jizzy 'tikiti-too' note.

WHITE-BELLIED SWIFTLET (Glossy Swiftlet)
Collocalia esculenta (10 cm)

The White-bellied Swiftlet is common everywhere, and is the only swiftlet that is easy to identify, by its glossy blue-black upper-parts and white belly. It breeds in dark rock shelters, bridges, and culverts.

The two swiftlets that are well known for their edible nests, and also their ability to navigate in darkness by echo-location, are the EDIBLE-NEST SWIFTLET *Aerodramus fuciphagus*, whose white nests are made of saliva, and the BLACK-NEST SWIFTLET *A. maximus* whose nests are mixed with vegetable matter. They nest mainly in caves in coastal areas, but forage widely inland.

White-bellied Swiftlet

SILVER-RUMPED SWIFT
Rhaphidura leucopygialis (12 cm)

The needletails are more heavy-bodied swifts, with butterknife-shaped wings (broader towards the tip) and short, rounded tails in which the feathers terminate in small spines. The Silver-rumped Swift is the smallest of the group and is a common resident in forested areas. It is glossy black with shining white rump and upper-tail coverts, and it is easy to identify.

Silver-rumped Swift

Among the larger needletails, the BROWN NEEDLETAIL *Hirundapus giganteus* (25 cm) is resident. These are the swifts whose passage recalls that of a jet plane: one hears the noise of the wings before seeing the birds! However, there are also two large migrant species and their identification is best left to the experts.

WHISKERED TREESWIFT

Hemiprocne comata (15 cm)

Whiskered Tree Swift

The treeswifts are not true swifts but are similar in flight, with long bodies and very long, slender wings. Whiskered Treeswifts habitually perch in trees in the forest or on the edge of clearings, and sally forth after insects, returning to perch and folding their wings in jerky movements, like soldiers on parade closing arms. At rest, the very slender wing tips cross each other above the long forked tail. Usually in twos or threes, sometimes they gather to roost in quite large numbers at favoured places. They are handsome and quite colourful birds, with a bronzy body, bluish head and wings, chestnut cheeks in the male, and two long white facial 'whiskers'.

GREY-RUMPED TREESWIFT

Hemiprocne longipennis (20 cm)

Grey-rumped Tree Swift

This treeswift spends longer in the air and thus wanders further from the forest. The very long, needle-forked tail is distinctive, and a large crest is visible when the bird is perched. The upper-parts are greenish and the under-parts are pale, and the male has chestnut cheeks.

Trogons

THE trogons are some of Malaysia's most colourful forest birds, but they tend to sit quietly in the middle storey of dense forest, so it is difficult to obtain good views. When disturbed, they will usually fly a short distance to a new perch, with just a tantalizing tip of the tail left to view. However, they are quite vocal, and their calls can be learnt. The wings are short and rounded, and the tail is moderately long and square tipped. The females are also colourful though they lack the very bright colours which serve to identify the males.

DIARD'S TROGON

Harpactes diardii (30 cm) Plate 8a

This is a characteristic trogon of dense lowland forest, in which the male has a black head and upper breast, a pinkish collar on the nape, and a narrow, crescentic pink band separating the black breast from the scarlet abdomen. The top of the crown is maroon. It is usually seen in pairs, but more often heard, a series of about eight to ten 'kau' notes falling slightly in pitch.

RED-NAPED TROGON

Harpactes kasumba (33 cm) Plate 8b

This very similar but slightly larger trogon occurs in the same habitat as the previous. The male has a more pronounced red hind collar and a white breast band. The call differs in that it consists of only six or seven slower notes on one pitch, but with care this difference can be learnt.

SCARLET-RUMPED TROGON

Harpactes duvaucelii (24 cm) Plate 8c

The Scarlet-rumped Trogon is smaller, and only the head of the male is black, both the breast and belly being scarlet, as is the rump. Being less confined to the interior of dense forest, it is more readily seen than the previous species, and it is also more vocal. The call is a cadence of some dozen notes beginning quite slowly and accelerating rapidly.

ORANGE-BREASTED TROGON
Harpactes oreskios (25 cm) Plate 8d

The mountain forests have a different set of trogons. Plate 8d shows this species to have a greenish head and an orange-yellow belly, or yellow in the female. The song is a five-note cadence.

Kingfishers

KINGFISHERS are colourful birds but many are shy and difficult to see, except as a dash of colour vanishing around the next bend in the river. There are at least a dozen species, and they occur in a range of habitats in both forest and open country, not always near water. Those living within the forest are especially difficult to see, as they will sit quietly through much of the day, in contrast to some of their noisy open-country cousins. The more commonly seen kingfishers are described below, beginning with the smallest.

RUFOUS-BACKED KINGFISHER
Ceyx rufidorsum (14 cm) Plate 9a

This tiny kingfisher, also known as Dwarf Kingfisher, can often be seen by small rivers and in swampy areas in the forest, usually as a little darting reddish ball with a long beak. There are two forms, generally treated as separate species, though hybrids occur. This redder form is more common, and is a lovely lilac-mauve colour, brighter on the rump, and the under-parts are yellow. The darker BLACK-BACKED KINGFISHER *C. erithacus* has a blackish mantle and wings; this one is also migratory and so may turn up in a variety of habitats. The call is a soft, high squeak.

COMMON KINGFISHER
Alcedo atthis (18 cm) Plate 9b

The Common Kingfisher occurs widely in winter throughout the region, in open swamps, on river banks, and along coasts. A few stay to breed in the northern states, occupying holes in the banks of rivers. The rufous ear coverts distinguish it from the resident BLUE-

EARED KINGFISHER *A. meninting* (15 cm), which is a much deeper blue above, has blue ear coverts, and is found along rivers in forested country.

RUFOUS-COLLARED KINGFISHER

Actenoides concretus (23 cm) Plate 9d

The true forest kingfishers that live away from pools or rivers are shy birds that are rarely seen, prone to sit motionless for long periods, and perhaps mainly active at dusk or dawn. The observer who has the good fortune to chance upon this species will see the lovely birds illustrated in Plate 9d. The call is reported to be a rapidly repeated rising whistle but is rarely heard.

Much more often heard is the call of the BANDED KINGFISHER *Lacedo pulchella* (21 cm), which is a forest canopy bird. The call is readily recognized, a series of disyllabic notes over about 15 seconds: 'taweeo, t'weeo, tu-teo, tu-teo, tu-teo', etc. The male plumage is heavily barred black on blue, while the female is barred black on chestnut.

COLLARED KINGFISHER

Halcyon chloris (24 cm) Plate 9c

Originally a coastal kingfisher, this common bird has spread inland, though it requires wooded country such as rubber estates. It has greenish-blue upper-parts with a white collar, and shining white under-parts. Casuarina trees along the beach often ring with its screaming and rather irritating 'kek-kek, kek-kek-kek' calls, and a slightly softer 'pit-prree, pit-prree'.

WHITE-THROATED KINGFISHER (W)

Halcyon smyrnensis (28 cm) Plate 9e

This is the big kingfisher that occurs widely in open country. The chestnut, white and blue plumage, with a red bill, is distinctive, and white patches are prominent on the wings in flight. The call is a loud and raucous, slow descending trill. It is by no means confined to water, although fish form part of the diet.

There is also a rather beautiful species that migrates from northern Asia from September to March, the BLACK-CAPPED KINGFISHER *H. pileata* (31 cm). Seen mainly along the larger rivers, it has a prominent black cap, white collar, and whitish under-parts that become rufous on the belly. However, when viewed in flight from the rear, it is very similar to the White-throated Kingfisher.

STORK-BILLED KINGFISHER
Pelargopsis capensis (37 cm) Plate 9f

The largest kingfisher is unmistakable, with its huge red bill, blue wings (without a white patch), and bright blue back. The head is brown and the under-parts buffy. Mainly a coastal bird, it also extends far up rivers and into open freshwater swamps. The calls are very loud and rather wailing, though not always unmusical.

Bee-eaters

BEE-EATERS are so named because some species include flying bees as a regular part of their diet, but they will take any flying insects. They are slender, graceful birds, and, like nightjars, swifts, and swallows, they are skilful aerial hunters. The bill is fine and slightly curved, and the wings are rather triangular in shape.

BLUE-THROATED BEE-EATER
Merops viridis (28 cm) Plate 11a

The Blue-throated Bee-eater is most often seen along logging tracks in well-wooded areas of the lowlands, hawking for insects at or below tree-top level in twos or threes, coming to rest periodically on exposed branches, but never really sitting quietly. The call is a soft 'kwilp, kwilp'. It nests in burrows in banks, singly or in groups, and availability of sandy banks defines where it is found in the breeding season. Thus coastal and riverine sites are especially favoured. Outside the breeding season, it is often seen in small flocks in more open country, and it migrates locally, with birds from the north replacing at least part of the breeding population during the northern winter.

It can be confused with the BLUE-TAILED BEE-EATER

M. philippinus (30 cm), which is a common migrant between August and April. This differs in having a green head, and a pale yellow throat bordered below by a brown gorget on the breast. Both species have the pale blue rump and tail. A few have bred in the northernmost states, where the CHESTNUT-HEADED BEE-EATER *M. leschenaulti* (22 cm) can be found locally, breeding in sand banks. This smaller bird also has a yellow throat bounded below by a narrow gorget, but it has a chocolate cap and back, green wings, and blue tail.

RED-BEARDED BEE-EATER
Nyctiornis amictus (33 cm) Plate 11b

This is a much more sluggish bee-eater that rarely leaves the forest. Plate 11b shows the prominent red throat and breast, and lilac forehead. The harsh call of four or five descending 'ka-ka-ka-ka' notes readily reveals its presence, usually perched in twos or threes in the middle storey.

Rollers

DOLLARBIRD
Eurystomus orientalis (30 cm) Plate 11c

Rollers are named from their rolling flight in courtship display. The Dollarbird or Broad-billed Roller is a rather solid dark bird, with red legs and thick red bill, and acquires its name from a circular, pale silvery-blue patch on the wings, visible in flight. It is seen in wooded areas and forest clearings, singly or in pairs on exposed perches, from which it watches for insects in the air, or occasionally on the ground. The voice is harsh and rasping.

Hornbills

EVEN people with no interest in birds will enquire about the hornbills and indeed they are very impressive, with their massive build, prominent bills, and loud calls. West Malaysia has nine species. They are broad-winged and long-tailed, and the larger species make quite long flights from one patch of forest to another in their search for trees in fruit. The wings of some species are very noisy, like the 'whoosh, whoosh' of a steam engine, often the first indication that hornbills are flying overhead. All the hornbills have a distinctively shaped horny protuberance or casque above the bill. They nest in holes in the trunks of trees, and seal up most of the entrance hole with mud, thus imprisoning the female during incubation, presumably as a protection against monkeys.

RHINOCEROS HORNBILL

Buceros rhinoceros (male 120 cm, female 90 cm) Plate 10c

This is perhaps the best known of the hornbills. It is a large, noisy bird, black, but with a white belly, and a white tail with a black bar across the middle. The up-turned casque is yellow in front, reddish behind. The call is a series of single loud honks, 'hok, hok', which usually breaks into a raucous, goose-like 'g-raak, g-raak' as the bird takes flight, audible from afar.

GREAT HORNBILL

Buceros bicornis (125 cm) Plate 10b

This magnificent bird resembles the previous species in plumage and voice, but is distinguished by its huge size, different shape of casque, and yellowish-white neck and wing bar. It is the latter that most readily identifies this hornbill in flight. In our region it is confined to the northern half of the peninsula, in both the lowlands and hills, recorded south to Selangor and Pahang.

PIED HORNBILL

Anthracoceros albirostris (75 cm)

This hornbill, the smallest, is the one most seen by those who do not customarily enter the forests, for it occurs in wooded areas and forest margins along rivers, swamps, and coasts. The figure at right shows the black and white plumage. The rear edge of the wings and the outer tail feathers are white, conspicuous

Pied Hornbill

in flight. Although the calls are not as loud as those of the previous species, they are shrill and vociferous birds with a variety of yelping calls: 'yip, yip, yip, ayip, youk, youk' and so on.

BLACK HORNBILL

Anthracoceros malayanus (75 cm)

The Black Hornbill is all black, except for the white tips to the outer tail feathers, and the bill and casque which are white in the male. The call is a coarse, rasping note that recalls the squealing of a nervous pig. Some males (illustrated) have a white stripe over the eye to the nape.

Black Hornbill

BUSHY-CRESTED HORNBILL

Anorrhinus galeritus (70 cm)

This hornbill lives in sometimes noisy flocks in or just beneath the canopy of the forest. It is all blackish except the basal two-thirds of the tail which is grey-brown. The calls consist of shrill yelping, similar to that of the Pied Hornbill but the notes are mostly disyllabic.

Bushy Crested Hornbill

WHITE-CROWNED HORNBILL

Berenicornis comatus (85 cm)

This distinctive long-tailed black and white hornbill lives mostly in the middle and upper storeys of the forest. It is a quieter bird with softer calls, and appears to be quite rare or local in distribution. The head and most of the under-parts and the tail are all white, with a bushy white crown, while most of the wings and the under-tail coverts are black. The female has black under-parts.

White-crowned Hornbill

WREATHED HORNBILL

Rhyticeros undulatus (100 cm) Plate 10d

The male Wreathed Hornbill is black with a white neck and tail, though these are often stained yellowish. The small crest is chestnut, and the casque is comparatively small. There is a pouch on the throat, yellow in the male and blue in the female, crossed by a narrow black band. The neck and under-parts of the female are black. It lives in the forested areas of the lowlands and lower hills. The call is a harsh barking note, but the bird is not very vocal. Its wings are very noisy in flight.

WRINKLED HORNBILL

Rhyticeros corrugatus (75 cm) Plate 10e

Although smaller than the Wreathed Hornbill, size is not a useful field indicator, and care is needed to distinguish the two species. The main features to look for are the lack of a black bar on the pouch, the differently shaped red casque, and the fact that only the terminal two-thirds of the tail are white (all the tail is white in the Wreathed). As always, the white feathers are often stained yellowish. Being a bird mainly of lowland and swamp forest, it is becoming rare.

HELMETED HORNBILL
Rhinoplax vigil (125 cm, female smaller, but both have at least an additional 20 cm of elongated central tail feathers) Plate 10a

This stunning bird, which is mainly found in the more hilly forests, is a favourite to anyone who has heard its call. It is huge and unmistakable, particularly with its long tail feathers. The extraordinary call begins with slow, single 'hoop' notes at long intervals, eventually speeding into disyllabic notes, 'te-hoop', which build to a crescendo, that then breaks into a cadence of 'poo' notes, abruptly terminated by a rapid, descending, cackling laugh. The whole (abbreviated) sequence, which may take several minutes, is as follows: 'hoop - -/ hoop - -', etc., 'te-hoop, te-hoop, te-hoop, poo, poo, poo, poo, ka-ka-ka-ka-ka'. Commonly the second bird will both begin and end its sequence shortly behind that of its mate. The neck is orange in the male and pale bluish in the female.

Barbets

THE main feature of many barbets is the relentless persistence of their calls, so much so that after a while an observer no longer notices them. The bills remain closed while calling, and the heads are turned, giving a ventriloquial effect. Barbets are stocky birds with heavy bills and short tails, and they are poor flyers, flying only short distances on audibly whirring wings. Most are predominantly green but with a startling array of gaudy colours in the head region, which are diagnostic when visible, but this is unusual as the birds are generally hidden in the canopy. The sexes are alike in most species. They are recognized mainly from their calls, although some species have two quite different calls, even three, which is confusing.

1. Lowland barbets

RED-CROWNED BARBET
Megalaima rafflesii (25 cm) Plate 11d

The Red-crowned Barbet is one of the characteristic barbets of the lowland forests. The call consists of two notes: 'tuk, tuk', a pause, and then some ten to twenty such notes delivered at constant speed for a few seconds. The pitch and speed of delivery may differ between individuals, but the pause after the second note is absolutely unvarying.

BLUE-EARED BARBET
Megalaima australis (18 cm) Plate 11e

Probably the commonest barbet in the lowlands and lower hills, the Blue-eared Barbet has also the most monotonous call, seemingly repeated for hours at a time. There are three types of call, the most typical being an endless disyllabic 'tr-trrk' repeated at about two per second. Another call is monosyllabic, uttered at the same speed or slightly faster, but for short periods, which has been likened to the sound of a football referee's whistle. A third call is a rather tinny rolling note, repeated at shorter intervals and quickening in speed.

GOLD-WHISKERED BARBET
Megalaima chrysopogon (30 cm) Plate 11f

The call of this species is a loud 'te-hoop', repeated rapidly for many minutes at a time. There is also a trilling call, beginning with a long trill that is repeated in gradually shortening phrases, until they break into a long series of triple and finally double notes. This species occurs mainly in more hilly terrain.

RED-THROATED BARBET
Megalaima mystacophanos (23 cm) Plate 12a

Another barbet of the lowlands and lower hills, this is not quite so common. The call of this species consists of 'tuk' notes uttered in irregular phrases, typically one, two, or three at a time, with short breaks. It also has a rolling or trilled note, repeated in shortening phrases, similar to that of the previous species. It is seen in Plate 12a that the sexes differ.

YELLOW-CROWNED BARBET

Megalaima henricii (22 cm) Plate 12b

This lowland barbet has an unmistakable call that consists of a roll and four 'tuk' notes, repeated: 'trrrrk-tuk-tuk-tuk-tuk, trrrrk-tuk-tuk-tuk-tuk', etc.

COPPERSMITH BARBET (W)

Megalaima haemacephala (15 cm) Plate 11g

The smallest barbet, this has a regular 'tonk tonk' call from which it gets its name. This is not a forest bird, being found in wooded areas and around villages, but it does not seem to be very common. Another lowland barbet, in lightly wooded country of the northern states, is the LINEATED BARBET (W) *M. lineata* (28 cm), which has a brown body, heavily streaked with white. The call is an emphatic 'bul-tuk', the second note sharper and a bit higher, repeated in phrases or at irregular intervals. Occasionally, it gives a series of more musical trilled notes, similar to those of the Gold-whiskered Barbet.

BROWN BARBET

Calorhamphus fuliginosus (18 cm) Plate 11h

As the different generic name implies, this bird is quite unlike the other barbets in all respects. It feeds in parties in the middle storey, it is brownish throughout, paler below, with conspicuous red feet, and its call is a forced, sibilant, thin whistle, uttered in short phrases. This call is inconspicuous, but immediately recognizable to those already familiar with it.

2. Mountain barbets

None of the birds described here occur in East Malaysia, which has its own endemic montane barbets (see Appendix 1).

BLACK-BROWED BARBET (W)
Megalaima oorti (20 cm) Plate 12d

The Black-browed Barbet is heard commonly throughout the mountains at 1000 or 2000 metres elevation. The call is a repeated, rapid three-note phrase, the second note softer and the third rolled: 'took-tk-trrrk', repeated about once per second.

FIRE-TUFTED BARBET (W)
Psilopogon pyrolophus (28 cm) Plate 12c

Another montane barbet, the Fire-tufted has an extraordinary cicada-like buzzing note, quite unlike the repetitive 'tooks' and 'poops' of the other species. The buzz begins as single notes which gather speed until ending in a longer, harsh buzz, the whole phrase lasting about three seconds.

GOLDEN-THROATED BARBET (W)
Megalaima franklinii (23 cm) Plate 12e

Confined to the higher mountains, above 1300 metres, this barbet has a di- or tri-syllabic note repeated at a rate of at least one per second, accented on the final syllable.

Woodpeckers

WOODPECKERS form a big family, with twenty-four species in our region. They are adapted to feed on insects and wood-borers hidden in crevices in the bark of trees, for which they have strong, pointed bills. Their stiff tails are used as a support while climbing vertical tree-trunks. They nest in holes, which they have bored or enlarged themselves. When feeding, the tapping of bill on wood will often attract attention, and some species 'drum' very rapidly with their bills as a form of territorial vocalization. They are quite strong flyers, with characteristic dipping flight on short, rounded, whirring wings. All are resident, and many species range widely in forests and woodland edges through the lowlands and hills. Most have either sharp, chittering or shrill call notes, often not very helpful for identification. It is difficult to make a selection from so many species, and only the commoner ones are described, divided into five groups.

1. Tiny woodpeckers (less than 10 cm): The Piculets

RUFOUS PICULET
Sasia abnormis (9 cm)

Although no bigger than a flowerpecker, the short tail and behaviour of this unobtrusive bird show that it is indeed a woodpecker. This pleasing little bird will usually reveal itself only by its persistent, quiet tapping. It is olive-green above and orange below, with a patch of gold or orange on the forehead. It occurs around decaying wood low in the forest and bamboos, mainly in the lowlands. The SPECKLED PICULET *Picumnus innominatus* (10 cm) of montane forests is strongly patterned with two white stripes in the head and bold black spots on the under-parts.

Rufous Piculet

2. Small woodpeckers (13–22 cm), having patterned or barred plumage with few bright colours, often found in lightly wooded terrain

BROWN-CAPPED WOODPECKER
Picoides (Dendrocopus) moluccensis (13 cm) Plate 13a

Also small and unobtrusive, this woodpecker occurs in wooded growth near the coasts, especially mangroves. Mainly brown and white, the head gives a striped appearance, and the male has a red streak in the cheek. It is seen on low bushes or the lower branches of trees, including village fruit plantations, where its soft but quite forceful chittering notes often reveal its presence.

GREY-AND-BUFF WOODPECKER
Hemicircus concretus (13 cm) Plate 13b

This more distinctly marked woodpecker has a pronounced triangular crest. The crest and forehead are crimson in the male, but grey in the female, the back has a striking scalloped pattern of black and buff, and the rump is white.

Two slightly larger species have strongly barred plumage. The BUFF-RUMPED WOODPECKER *Meiglyptes tristis* (19 cm) is dark with strong whitish bars and rump, and a red cheek patch in the male. It has a triangular crest but this is not often raised. The BUFF-NECKED WOODPECKER *M. tukki* (22 cm) is more finely barred, with a pale buff patch on the sides of the neck, but not on the rump. Again the male has a red cheek patch, but the crest is not prominent. All three species are quite common in lightly wooded country and forest edges. They often perch across the branches, and all have shrill chittering or short trilling calls.

3. Medium woodpeckers (23–27 cm) without green in the plumage

BAY WOODPECKER (W)
Blythipicus pyrrhotis (27 cm) Plate 13c

This is a rather large woodpecker of the hill stations, identified by its brown-maroon plumage, barred black on the upper-parts, and its large yellow bill. Only the male has the red neck patches.

In the lowlands, it is replaced by the smaller MAROON WOODPECKER *B. rubiginosus* (23 cm), which also has a yellow bill, but the plumage is unbarred. However, the RUFOUS WOODPECKER *Celeus brachyurus* (25 cm) is a rather similar lowland species, being dark rufous with some black barring, but with a dark bill; the male has a red cheek patch. Both are quite common in a range of forested and wooded habitats.

4. Medium woodpeckers (25–34 cm), having green plumage with extensive red and yellow, belonging to the genus *Picus*

CRIMSON-WINGED WOODPECKER
Picus puniceus (25 cm) Plate 13d

One of the three common green woodpeckers of lowland forests, the Crimson-winged Woodpecker has a green head with red crest, tipped yellow, and red wings.

The CHECKER-THROATED WOODPECKER *P. mentalis* (28 cm) has a yellow crest, and no red on the head. It has a reddish-chestnut collar and wings, and a black-and-white checkered thoat. The BANDED WOODPECKER *P. miniaceus* (25 cm) has an all red head, with yellow-tipped red crest, and strong barring especially on the under-parts.

All three have similar calls, variously described as 'shay', 'sheok', or 'queer-loo', either singly or in runs of a few notes. It is difficult to separate them by their calls, perhaps because these are so often uttered at dawn or dusk when the bird is silhouetted at the top of a tree and the colours cannot be seen.

GREATER YELLOWNAPE (W)
Picus flavinuchus (34 cm) Plate 13e

This and the next species replace the previous three in the mountains, at elevations above 1000 metres although there is some overlap. This species is plain green above and below, except for its yellow crest and throat, and banded wings. At slightly higher elevations, the LESSER YELLOWNAPE (W) *P. chlorolophus* (27 cm) differs in having some red and a white stripe on the front of the head and a barred belly.

5. Larger woodpeckers

COMMON GOLDENBACK
Dinopium javanense (30 cm) Plate 13f

The Goldenback or Flameback is a large, prominently coloured woodpecker of secondary growth and gardens, including rubber and coconut plantations. With its golden-yellow wings and dipping flight, it might be mistaken for an oriole, but it has a more elongated form, in addition to having a red crest (in the male) and rump, banded black and white cheeks, and white under-parts with black crescentic markings.

The ORANGE-BACKED WOODPECKER *Reinwardtipicus validus* (30 cm) is a handsome bird, characterized by a long buff stripe from the back of the neck to the rump, giving an elongated, long-necked appearance. The under-parts and crest are red, and there are chestnut bars in the black wing. It has a typical woodpecker chittering call.

WHITE-BELLIED WOODPECKER
Dryocopus javensis (43 cm)

This is a large woodpecker of the lowland forests, a great black bird with a red crest (in the male) and white belly. Its call is a very loud, metallic 'sheok' that rings through the forest.

Finally, the largest woodpecker, the GREAT SLATY *Mulleripicus pulverulentus* (51 cm), is easily identified by its size, long-necked appearance and slaty colour, with a buff throat. The loud whickering or whinnying call, 'wooik, wooik' and 'wi-wi-wik', can often be

White-bellied Woodpecker

heard in lowland forest, perhaps followed by a view of three or four in flight at canopy height, with large, rounded, dark wings.

Broadbills

THE broadbills are a colourful family of arboreal birds, with rather squat bodies, and bills that seem almost ridiculously broad if viewed head-on. Most broadbills are heard far more often than seen, and one or two species will be heard on any day spent in the forest. Once the songs have been learnt, they can be recognized immediately. There are seven species in West Malaysia.

BLACK-AND-YELLOW BROADBILL

Eurylaimus ochromalus (15 cm) Plate 14a

The Black-and-Yellow is the broadbill that is most commonly heard in the forests of the lowlands and lower hills. The loud song begins slowly and gathers speed until it reaches a cicada-like buzz, ending abruptly. The plumage, which is the same in both sexes, is very striking, but it is very difficult to obtain a good view. Both sexes have the same song, usually one of a pair beginning its song slightly behind its mate, or uttering plaintive 'piu' calls in accompaniment.

BANDED BROADBILL

Eurylaimus javanicus (23 cm) Plate 14b

The Banded Broadbill's song is very similar to that of the previous species, except that it begins with a short flourish, followed instantly by the cicada-like series of rapid notes uttered at constant speed, until the final notes which tail off at the end. The female lacks the dark breast band shown in Plate 14b. The bill is all blue.

BLACK-AND-RED BROADBILL

Cymbirhynchus macrorhynchus (25 cm) Plate 14d

This broadbill occurs mainly in riverine forests, and is often seen as it flies across the rivers. The plumage is similar to that of the Banded Broadbill, but it has white in the wing, and the bill is blue above and yellow below. It is a much more silent bird, occasionally giving a quiet but quite harsh note.

DUSKY BROADBILL

Corydon sumatranus (28 cm) Plate 14c

The largest of the lowland broadbills, this species is dark in colour, with a pale buff throat and a pinkish bill. The song consists of the repetition of a rather lisping 'tiu-wit', four or five times, which slightly ascends the scale.

GREEN BROADBILL

Calyptomena viridis (19 cm) Plate 14f

Being all green, this is not an easy bird to see in the foliage of the canopy. As Plate 14f shows, it is a curiously shaped bird, very chunky with the feathers of the forehead partly covering the bill. The female lacks the black in the neck and wings. Its voice is heard commonly in the forests of the lowlands and lower hills, and can be recognized instantly once known: a short bubbled note, emphasized at the start and with an upward inflexion, having the timbre of a stone bouncing on glass. There are other calls but this is the most typical.

LONG-TAILED BROADBILL

Psarisomus dalhousiae (28 cm) Plate 14e

This rather local, montane broadbill is mainly green, offset by brilliant yellow about the face and throat, and blue in the wings and on the long tail. The crown is black. The song is a typical broadbill run of notes, beginning suddenly and gathering speed.

The SILVER-BREASTED BROADBILL (W) *Serilophus lunatus* (18 cm) of the mountains is an ashy coloured bird, with a blue bill, chestnut rump, black tail, a blue patch in the wings, and a broad black stripe above the eye. In flight, it has a white patch in the wings and white tips to the tail feathers. The female has a narrow silver band across the breast.

Pittas

THIS is an enigmatic family. Although often having brilliant colours, they are secretive birds of the forest floor, and the casual bird-watcher is lucky if he chances upon one. They are stocky birds, with very short tails, short and rounded wings, and rather long legs. The calls are distinctive, and the best way to see a pitta is to identify the call, and then to stealthily follow it, but this requires time and patience. Despite their secretive behaviour, some species are migratory. There are seven species in West Malaysia.

GARNET PITTA
Pitta granatina (18 cm) Plate 15a

Although the plumage shown in Plate 15a is distinctive, the Garnet Pitta is a very elusive bird. However its call, consisting of a soft monotone whistle of slightly over one second duration, with a slight upward inflexion, will often reveal its presence. Unfortunately, another ground frequenting bird in the forest, the Rail-babbler (see p. 64), has an almost identical call, distinguishable only to the practised ear. It is mainly a lowland bird.

The BANDED PITTA *P. guajana* (23 cm), which occurs in lowland forest, has a short 'pow' note and a low 'crrrrr', which can be easily overlooked. It has broad black and orange-yellow stripes on the head, white in the wings, and a finely banded blackish and orange abdomen. The tail is blue.

BLUE-WINGED PITTA
Pitta moluccensis (20 cm) Plate 15b

Plate 15b shows another colourful bird. In flight away from the observer, it appears mainly blue, with a large white patch in the wings. The call is very distinctive, a double whistle best rendered as 'hoyuk-oyuk'. Mainly a migrant to our region, to a variety of scrubby habitats, it breeds in the forest margins on Langkawi Island and in the north-west states. The situation is confused by the presence of an almost identical bird, the MANGROVE PITTA (W) *P. megarhyncha*, which is resident but confined to mangroves.

Swallows

THE swallows are well known to most people, being familiar birds of towns and villages. Swallows are superficially like swifts, but the wings are not so long and crescentic, and the flight is different. They are largely aerial, hawking for insects with characteristic swooping flight. They roost commonly along electricity and telegraph wires in down-town areas, and some roosts number many thousands. There are two common species, both more or less blue above and whitish below.

PACIFIC SWALLOW
Hirundo tahitica (14 cm)

This is the resident swallow, with a pair or two found in every village, usually building a mud nest on the rafters or eaves of houses. It has a chestnut throat, and the belly is rather dirty white. The outer tail feathers are slightly elongated.

The BARN SWALLOW *H. rustica* of Europe and Asia is an abundant winter migrant. It greatly outnumbers the resident birds from about August to May, and a fair number remain through the summer. The adult has more elongated outer tail feathers, and a cleaner white belly, which is separated from the chestnut throat by a black band. However, many wintering birds are immature, and the two species are then more difficult to distinguish.

Pacific Swallow

RED-RUMPED SWALLOW
Hirundo daurica (18 cm, excluding elongated outer tail feathers)

Plate 15c

This larger swallow, with a deeply forked tail and slower, soaring flight on broader wings, is so named from the rump which varies in colour from pale rufous to chestnut according to subspecies. It is sometimes also known as Striated Swallow *H. striolata*, from the ventral striations, but the name used here is preferred for birds in Malaysia. The resident form (subspecies *badia*) illustrated in Plate 15c is particularly dark, with all chestnut under-parts. It is seen around limestone crags. Much paler forms, with more heavily streaked under-parts, are scarce winter migrants.

Wagtails and Pipits

GREY WAGTAIL
Motacilla cinerea (19 cm)

The wagtails, so named from their long tails which are constantly wagged up and down while they feed on the ground, are common winter migrants to our region, arriving in July or August. The Grey Wagtail has grey upper-parts and yellowish under-parts. In winter, the throat is white, though it becomes black in the breeding male. The call is a sharp, double 'chizik'. The normal habitat of Grey Wagtails is rocky mountain streams, but in winter they are usually found along tracks, especially in the hills.

In contrast, the YELLOW WAGTAIL *M. flava* (18 cm) is a common migrant in small flocks to rice-fields and open swamp areas. It has a rather shorter tail and more olive green upper-parts, and the call is a single, wheezy 'tseep'.

RICHARD'S PIPIT *Anthus novaeseelandiae* (18 cm) is a ground bird of open grassy terrain. Brownish and slightly streaked, it is a slender bird with quite long legs and, like the wagtails, it has white outer tail feathers. It is not uncommon in a variety of habitats, including airports, but easily overlooked by bird watchers interested in seeking the more colourful birds of the forests.

Grey Wagtail

Cuckoo-shrikes

BAR-WINGED FLYCATCHER-SHRIKE
Hemipus picatus (15 cm)

Bar-winged Flycatcher-shrikes usually occur in small parties in the upper canopy, where they are easily overlooked, though their pleasant but rather indistinctive call will often reveal their presence. Identification is then not difficult, but note that another species, the BLACK-WINGED FLYCATCHER-SHRIKE *H. hirundinaceus*, is identical except that it lacks the prominent white wing bars. The females of both species have the black replaced by brown. The Bar-winged occurs mainly in the hills, the Black-winged in the lowlands.

Bar-winged Flycatcher Shrike

PIED TRILLER
Lalage nigra (18 cm)

The other black-and-white member of the group is a more familiar bird of the lower and middle canopy in wooded areas, including plantations, gardens, and mangroves. The plumage is quite distinctive, with the black again replaced by brownish-grey in the female. The call is a short, slow trilling note on one pitch, with a timbre not unlike the chirp of a sparrow.

Pied Triller

MALAYSIAN CUCKOO-SHRIKE (W)

Coracina javensis (28 cm)

The true cuckoo-shrikes are larger birds of varying shades of grey, and their taxonomy is rather confused. Being birds of the canopy, they are conspicuous only when they come out from dense forest, although their loud calls often reveal their presence. This bird of the mountains is medium grey, darker on the face and very pale on the belly.

The lowland species of cuckoo-shrike are the LESSER CUCKOO-SHRIKE *C. fimbriata* (19 cm), which is smaller and darker, and the BAR-BELLIED CUCKOO-SHRIKE *C. striata* (28 cm), which is medium grey without a dark mask, and with pronounced barring in the female on the under-parts and rump.

Malaysian Cuckoo Shrike

Minivets

LIKE the cuckoo-shrikes, to which they are related, the minivets are wholly arboreal. They are delightful birds with their long tails and brilliant plumage, commonly red in the male and yellow in the female. They characteristically occur in parties of up to a dozen or more birds, which keep together with constant twittering calls, but the parties may be loosely strung, with a tendency for individuals to fly from tree to tree in twos or threes.

SCARLET MINIVET

Pericrocotus flammeus (19 cm) Plate 15e

The glossy black and scarlet of the male, replaced by grey and yellow in the female, might be considered distinctive, but there are two very similar species. The Scarlet Minivet is a lowland bird, replaced in the mountains above 1000 metres by the GREY-CHINNED MINIVET *P. solaris*. However, in Malaysia, the grey chin of the latter bird is very dark, and altitude is perhaps the best distinguishing character; the alternative name of Mountain Minivet may be more appropriate.

The lowlands also have the smaller FIERY MINIVET *P. igneus* (15 cm). Size is an unreliable guide, so care is needed. In the male, the red has a fiery orange shade, while in the female, the rump, wing patch, and outer tail feathers are orange instead of yellow. It is safer to rely on the female for identification.

Quite a common winter migrant is the ASHY MINIVET *P. divaricatus* (20 cm). This is a grey bird, lacking any colours. The male is grey above and white below, with a white forehead and black hind-crown and nape, while the female is more drab. Both sexes show a white band in the wing in flight. It occurs in large parties in the tops of trees, mainly in the coastal lowlands.

Bulbuls

THE bulbuls form one of the principal families of smaller birds, with no less than twenty-four species in West Malaysia. They are quite noisy birds, usually found in the forest edge or canopy, and are characterized by soft and rather fluffy plumage. While most are forest birds, a few are found only in open country. Some bulbuls have distinctive plumage, but many are very difficult to identify, particularly the brown forest bulbuls in the genus *Pycnonotus*, and the more olive forest bulbuls in the genus *Criniger*. Only a representative selection is described here.

YELLOW-VENTED BULBUL

Pycnonotus goiavier (20 cm) Plate 16a

This common bulbul of the open country is a very unprepossessing species, with its dowdy plumage and rather thin though cheerful song. It is easily identified by the white stripes above and below the eye, and yellow under-tail coverts.

In Kedah, Perak, and Trengganu, another open country species is the RED-WHISKERED BULBUL (W) *P. jocosus* (20 cm), which is a cheeky and popular cage-bird that has become feral around some cities. Its black-and-white head region with tall black crest, white cheeks with a red patch, and white under-parts with red under the tail, are familiar to many people.

STRAW-HEADED BULBUL

Pycnonotus zeylanicus (29 cm) Plate 16b

The large Straw-headed Bulbul is renowned for its remarkable loud and musical song, though the song phrases lack variety. It is found in wet thickets along rivers through forest and in more open habitats. Being highly valued as a cage-bird, it has become almost extinct in the wild in Indonesia, and we should be careful to protect our own wild population.

BLACK-HEADED BULBUL

Pycnonotus atriceps (18 cm) Plate 16d

Among the forest bulbuls, the neatly plumaged Black-headed Bulbul is common generally. Important features to look for in this bird are the yellow tip to the black tail, and the yellow lower back. It has a pleasant, musical song, though the phrases do not vary.

BLACK-CRESTED BULBUL

Pycnonotus melanicterus (19 cm) Plate 16c

Occurring in the same habitat as the previous species, though usually at higher altitudes, this bulbul differs in having a crest, and lacking the yellow in the wings and tail.

SCALY-BREASTED BULBUL

Pycnonotus squamatus (14 cm) Plate 16e

There are two other forest bulbuls with very striking plumage, but being birds of the canopy, they are seen only rarely. The Scaly-breasted Bulbul shown in Plate 16e is unmistakable, and is found mainly in the hills.

The GREY-BELLIED BULBUL *P. cyaniventris* (16 cm) has a dark grey head and under-parts, yellow under the tail, and olive-yellow upper-parts.

RED-EYED BULBUL

Pycnonotus brunneus (18 cm) Plate 16f

The Red-eyed Bulbul is representative of the group of brown forest bulbuls which singularly lack distinctive features. At least two other species are almost identical, and even the colour of the eye-ring, from which two of the species get their names, is not a reliable feature. Identification is best left to the experts. The brown bulbuls have short but quite sweet and musical songs, which can be the key to their identity.

YELLOW-BELLIED BULBUL

Criniger phaeocephalus (20 cm) Plate 16g

The more olive-coloured bulbuls of the forest are usually slightly larger, and harsh of voice. The Yellow-bellied Bulbul is one of the easier species to identify, with its clean white throat and yellow under-parts. The head is grey and lacks a crest. Two similar species have rather duller plumage, but both sport a short crest, and both have notably harsh songs. The *Criniger* bulbuls will often puff out their white throat feathers in display.

Ioras, Leafbirds, and Fairy Bluebirds

THE ioras and leafbirds are mainly green birds that live in the foliage of the canopy, in woodland and forest.

COMMON IORA
Aegithina tiphia (15 cm) Plate 15d

This small bird is common in a variety of lightly wooded habitats, including mangroves, and is best known by its call, typically 'weeeeeeee-tu' and variations. Plate 15d shows distinctive plumage, but the GREEN IORA *A. viridissima* (14 cm) is rather similar. This is dark green both above and below. The male has a small yellow patch above and below the eye, and the wing bars of the female are pale yellow. However, the Green Iora tends to keep to the forest canopy and is more difficult to observe.

BLUE-WINGED LEAFBIRD
Chloropsis cochinchinensis (18 cm) Plate 15f

As the name implies, the green leafbirds are readily camouflaged in the canopy where they feed. The three species of the lowlands are very similar, all the males having black throats narrowly bordered by yellow, while the females are wholly green with a yellow wash. The light blue patches on the wings and tail of both sexes of this species render it the most distinctive of the three.

The GREATER GREEN LEAFBIRD *C. sonnerati* (20 cm) and LESSER GREEN LEAFBIRD *C. cyanopogon* (17 cm) lack this useful feature, and size alone is an unreliable field characteristic. The yellow throat of the female Greater Green Leafbird is a supportive clue when it is visible. All three have quite liquid, musical song phrases. Much more distinctive is the ORANGE-BELLIED LEAFBIRD (W) *C. hardwickii* (19 cm) of the mountains. In the male, the black throat is more extensive and the belly is orange, while the tail is dark, and there is blue in the wing. The female also has an orange wash in the belly.

FAIRY BLUEBIRD
Irena puella (25 cm) Plate 15g

This beautiful bird is quite common in lowland forests. The shining blue and velvety black male is difficult to mistake for any other bird. The female is more slaty blue with dark wings. A canopy bird, it is active and readily seen, and quite noisy with distinctive loud and mellow notes.

Shrikes

LONG-TAILED SHRIKE
Lanius schach (25 cm)

Shrikes are medium-sized birds with strong heads and hooked bills used for catching large insects, which are sometimes impaled on thorns for storage. This is a rather large and long-tailed species that is a common resident in open country. The upper-parts are grey, black on the forehead and through the eye, and the under-parts are white and chestnut-buff. Although the usual call notes are quite harsh, its short song is not unmusical.

The smaller BROWN SHRIKE *L. cristatus* (19 cm) is a common winter visitor to lightly wooded country. Slightly rufous brown above and creamy below, it has a broad black band through the eye, and a white stripe above it and on the forehead. Young birds show some fine crescentic barring. The call is a harsh scolding chatter.

Long-tailed Shrike

Thrushes

THE thrushes form a large family of insectivorous birds, mostly living in woodland, that includes the true thrushes, robins, and forktails. They often have quite striking plumage, and some are renowned songsters.

WHITE-RUMPED SHAMA

Copsychus malabaricus (27 cm) Plate 17a

Not only does the shama have beautiful plumage, it is also considered to be the forest's best songbird. The song consists of rich, liquid, warbled phrases, which can be quite varied, although singing ability seems to vary between individuals. However, it is a skulking bird of the undergrowth and not very easy to see, except for its white back as it flies into cover, where it will sit giving clicking notes in alarm, flicking its tail in accompaniment. The female has broadly the same plumage but is somewhat duller. It is found in the lowlands and hills throughout the region.

The RUFOUS-TAILED SHAMA *C. pyrropyga* (21 cm) is found in the lowland forests, more particularly perhaps the poorer forests on sandy or acid soils. The black is replaced by dark grey, with a prominent white streak in front of the eye, while the rump, base of the tail, and belly are rufous. The song is totally different, consisting of a series of singularly mournful glissading whistles.

MAGPIE ROBIN

Copsychus saularis (22 cm)

The black-and-white Magpie Robin is common in more open woodland and around villages. It is usually seen on the ground or lower branches, where it frequently cocks its tail. The song consists of short, sweet phrases that lack the fullness of the Shama's.

Magpie Robin

CHESTNUT-NAPED FORKTAIL

Enicurus ruficapillus (20 cm) Plate 17b

The forktails are named from their scissor-like, deeply forked, black-and-white barred tails. This species is the most colourful, with its chestnut nape and upper back. It lives along clear, swift and more rocky streams in the forest, especially in the hills but extending into the adjacent lowlands. Forktails in general are quite wary, and most often seen in flight along the river, but their shrill, whistled calls can be heard above the bubbling of the water. They are very active, dainty feeders, rather after the fashion of wagtails.

The WHITE-CROWNED FORKTAIL *E. leschenaulti* (25 cm) is black-and-white with a white crown, and is found in wet lowland forests, where its shrill call will reveal its presence.

BLUE ROCK-THRUSH

Monticola solitarius (22 cm) Plate 17c

As their name implies, these thrushes are typically associated with rocks, but they often occur around buildings, and in winter quarters also on coasts and in open fields. Most migrant birds, illustrated in Plate 17c, are blue with a chestnut belly, but an all blue resident form is found in the northern states, particularly around limestone outcrops. The bill is dark. Quite conspicuous birds, they have a loud and quite melodious thrush song at their breeding sites.

This bird should not be confused with the larger Whistling Thrushes which are also mainly blue, with varying shades of brown or purple. They are, however, very shy birds of the forest floor, also associated with rock outcrops and with running water. Usually only their alarm is heard, a single harsh whistle. The two species in West Malaysia have more or less yellow bills: these are the BLUE WHISTLING THRUSH (W) *Myiophonus caeruleus* (32 cm), a bird of lowland and lower hill forests, and the smaller MALAYAN WHISTLING THRUSH (W) *M. robinsoni* (25 cm), which is endemic to the mountains of the central states.

Babblers

THIS is the second big family of small forest birds, with some forty-seven species in our region. While some are easy to identify from their plumage, others lack distinctive features and are at least as difficult as the brown bulbuls. However, nearly all have distinctive songs or calls, although it requires intensive study to learn these. Published tape recordings of babbler songs may help considerably, if used with care. Identification is not assisted by the babblers' skulking habits, or the tendency of some groups to associate in mixed feeding parties. Space can be given here for only some of the more distinctive, common and representative species.

BLACK-CAPPED BABBLER
Pellorneum capistratum (18 cm) Plate 18b

This short-tailed babbler is common in the ground layer of dryland forest in the plains and lower hills. A very skulking bird, it walks stealthily about on the ground, but a quiet bird-watcher can usually obtain a view by following up the frequently uttered calls. As Plate 18b shows, the plumage is distinctive. The black cap appears as a black stripe along the crown. The typical call is a soft, slightly disyllabic rising whistle, 'pheet', but there are other calls, a common one being a double whistled 'tip-ee', the second note usually higher, though not always.

WHITE-CHESTED BABBLER
Trichastoma rostratum (16 cm) Plate 18a

The babblers in this genus look very alike, being various shades of brown above and creamy below, while the tails are rather short. They are skulking birds of the undergrowth of lowland primary or secondary forest. The White-chested Babbler favours dense water-side growth along rivers and streams, in swamps, and at the inland margin of mangroves. The rather loud song consists of three sliding or disyllabic whistles, the second pitched lower though the whole phrase rises in pitch. This is a cheerful song, in contrast to the very similar but less inflected and more monotonous song of the ABBOTT'S BABBLER

T. abbotti. The latter species is usually found in drier thickets of secondary growth.

RUFOUS-CROWNED BABBLER

Malacopteron magnum (18 cm) Plate 18c

Babblers of this genus are usually found in family parties in the undergrowth or lower storey of the forest. This is one of two species that have a dark rufous forecrown. The song is quite musical, consisting of clear whistled notes descending the scale, but sometimes rising at the start and end of the cadence, forming quite a complex sequence.

Very similar in appearance is the SCALY-CROWNED BABBLER *M. cinereum* whose song consists simply of five or six thin notes rising the scale.

A favourite and rather startling song of stream-side campsites in lowland forest is that of the SOOTY-CAPPED BABBLER *M. affine*, which heralds the dawn with its extraordinary tuneless whistle of single notes, pitched higher and lower in a sort of sequence, but never quite hitting the top notes. This has been likened to the aimless whistling of an errand boy.

LARGE WREN-BABBLER (W)

Napothera macrodactyla (19 cm) Plate 18d

The wren-babblers are so named because of their squat bodies with very short tails. Typically, they have rather mottled or scaly plumage, and they live close to the ground in a variety of forest habitats. The Large Wren-babbler is a rather uncommon bird of lowland forests. The song is usually a succession of complex phrases of loud, clear, and often slurred, whistled notes, the phrases repeated with variations introduced one at a time.

Most common among the montane forms is the STREAKED WREN-BABBLER (W) *N. brevicaudata* (14 cm), which has a long, shrill two- or three-note whistle.

BLACK-THROATED BABBLER

Stachyris nigricollis (15 cm) Plate 18e

The babblers of the genus *Stachyris* are quite distinctive in both plumage and song, and they are less prone than the other babblers to mixed flocking behaviour. The Black-throated Babbler is common in the lowlands, and has a striking song of clear, deliberate whistles on one pitch, a dozen or more notes of which the first one or two are emphasized. Typical of babblers, this song is frequently accompanied by low churring notes from the mate.

The CHESTNUT-WINGED BABBLER *S. erythroptera* is a very common bird of the undergrowth, and has a whistled trill on an even or slightly falling pitch; the trill is just too fast for a human to imitate, although attempts will usually incite a response. Distinctive features are the chestnut wings and dark fore-parts, with a blue eye-ring.

The larger CHESTNUT-RUMPED BABBLER *S. maculata* also has a blue eye-ring, but with a black throat, black spots on a white breast, and a chestnut rump. The song consists of clear di- or tri-syllabic phrases repeated in series, accompanied by trills from the mate. This song is rather easy to imitate, and the author has succeeded in whistling in this bird over a range of 100 metres. It is quite common in lowland forests.

The GOLDEN BABBLER (W) *S. chrysaea* (13 cm) is a smaller member of the group. Bright yellowish in colour, with a dark striped crown and a dark patch in front of the eye, it is very common in the mountains. It has a short whistled trill on one note, thinner and higher pitched than the Black-throated Babbler, and characterized by the emphasized first note, and usually a brief pause after the first or second.

FLUFFY-BACKED TIT-BABBLER

Macronous ptilosus (16 cm) Plate 18g

The Fluffy-backed Tit-babbler is named from the white-shafted spines on the back, although these can rarely be seen well. The most striking feature is the strange call, reflected in the Malay name. The male gives three or four hollow whistles, 'poop, poop, poop', which draws an immediate outburst of loud squawking from the female, 'kedeke-kreuer, kedeke-kreuer'. It is typically found in the secondary scrub and fern tangles of old ladangs or the margins of logging roads in lowland forest.

Another babbler of secondary growth is the STRIPED TIT-BABBLER *M. gularis* (13 cm), marked by a rufous crown and black streaks on the yellow breast. The call is a prominent and repetitive 'chonk, chonk, chonk', heard commonly in this habitat, and usually accompanied by chuckling notes from other members of the party.

CHESTNUT-BACKED SCIMITAR-BABBLER

Pomatorhinus montanus (20 cm) Plate 18f

This distinctive babbler with its curved yellow bill is a secretive bird, and thus appears to be uncommon. It is most often identified by its call, which consists of a mellow double or triple hoot. It is found in both lowland and hill forests.

BLACK LAUGHINGTHRUSH (W)

Garrulax lugubris (26 cm) Plate 19a

The charm of laughingthrushes lies in their light, hopping and inquisitive gait, and their outbursts of loud and cheerful noise. The family as a whole is one of the author's favourite groups of birds, but in our region they are confined to the hill forests. This species is all blackish, with an orange bill, and bluish bare skin around the face. It has a variety of calls, ranging from loud single hoots to a series of such notes followed by a maniacal laugh. It is found in the mountains above 850 metres.

CHESTNUT-CAPPED LAUGHINGTHRUSH

Garrulax mitratus (25 cm) Plate 19b

The Chestnut-capped Laughingthrush occurs commonly in flocks in the mountains. It is another attractive bird, with a prominent white stripe in the wing, and an orange bill and legs. The calls consist of three or four rich, thrush-like notes.

SILVER-EARED MESIA (W)

Leiothrix argentauris (19 cm) Plate 19c

This very colourful babbler is quite common in active and sometimes rather noisy parties in forest edge habitats in the mountains above 1000 metres. The plumage shown in Plate 19 is both beautiful and striking. The calls are varied, including a loud, cheerful song, but most often heard is a short, high, shrill chatter.

CUTIA (W)

Cutia niplanensis (18 cm) Plate 19d

The Cutia is a lovely bird that is active in the higher mountains, usually in pairs around the trunks of old trees in forest edge habitats, behaving rather like nuthatches (see page 71). It has twittering calls but no real song. The female has an olive-brown back marked with large black spots.

BLUE-WINGED SIVA (W)

Minla cyanoptera (16 cm) Plate 19e

Another locally common hill forest bird, this is brown above and white below, with a rather indistinct purplish-blue on the wings and tail, and pronounced stripes above the eye. The call is a triple slurred whistle.

WHITE-BROWED SHRIKE-BABBLER

Pteruthius flaviscapis (14 cm) Plate 19f

This is another colourful babbler of the mountain forests, found in the middle and upper storey. The female is less striking, with a grey head, faint eye-stripe, and lack of pronounced colours in the wing. It is quite common, occurring up to the moss forest level.

The LONG-TAILED SIBIA (W) *Heterophasia picaoides* (30 cm) is another montane babbler, found in flocks in the canopy. It is a long-tailed bird, grey or brownish-grey, with a white patch in the wing, and a paler tip to the blackish tail.

BROWN FULVETTA

Alcippe brunneicauda (15 cm) Plate 19g

The fulvettas or nun-babblers are rather nondescript, relatively large-headed brown babblers, that live in the middle storey of the forest and regularly participate in mixed feeding flocks. Although Plate 19g shows a singular lack of distinctive features, this bird is common in the lowland forests, and its rapid, tinkling cadence song soon reveals its presence.

MALAYSIAN RAIL-BABBLER

Eupetes macrocerus (29 cm) Plate 19h

This secretive bird of the forest floor was once thought to be very rare, until ornithologists learnt its call, since when it has been found to be quite common. It is a beautiful bird, as it walks purposely about on the ground, usually keeping just out of sight. However, imitations of its whistle will often produce a brief sighting, when one may see it flicking its wing as it walks, uttering soft, 'tuk-tuk' notes. The call is a plaintive, soft, monotone whistle of about a second or longer, which is easy to imitate, but it takes an expert ear to distinguish this from the almost identical call of the Garnet Pitta (see p. 47)—the pitta's call is lower, shorter, and sharper, with a slight inflexion. The Malaysian Rail-babbler is found in dryland forest in the plains and at least the lower mountains.

Warblers

THE warblers form a large family of rather small, thin-billed insect-eaters, most of them lacking bright colours. They are very active birds in the foliage, and named from their warbling songs, though most Malaysian species are not good songsters.

DARK-NECKED TAILORBIRD

Orthotomus atrogularis (11 cm) Plate 17d

The tailorbirds are small, long-tailed warblers that are quite noisy but secretive birds of the undergrowth and scrub. They are named from their remarkable tailored nests, which are constructed by stitching together one or more leaves to form a cup. Most tailorbirds are dull greenish above, and paler below. This species has a pronounced chestnut cap, green tail, and yellow under-tail coverts; there is a black throat patch, though this may not always be obvious. It is a common bird of forest margins in the lowlands and the lower hills, often seen alongside logging trails.

COMMON TAILORBIRD (W)
Orthotomus sutorius (10 cm) Plate 17e

There are three open country species in the lowlands. Most widely distributed is the Common Tailorbird which has a rufous forecrown, and the male in breeding plumage is distinguished by its elongated tail. The most distinctly marked tailorbird is the RUFOUS-TAILED TAILORBIRD *O. sericeus*, which has both the crown and tail rufous, and a clean whitish throat and breast. These two species have overlapping habitats, but the OLIVE-BACKED TAILORBIRD *O. sepium*, which has the whole head rufous, with a greenish tail, is more confined to coastal areas, especially mangroves.

All the above four species have monotonous mono- or di-syllabic calls, which it is useful to study. In contrast, the MOUNTAIN TAILORBIRD *O. cuculatus* has a varied musical repertoire. This has a longer bill, long white eye-brow, and bright yellow belly, and is found in thickets on the edge of montane forest.

YELLOW-BELLIED PRINIA
Prinia flaviventris (13 cm) Plate 17f

Superficially similar to tailorbirds, prinias have a more elongated appearance, with rather longer, slightly graduated tails, and more uniform plumage. The Yellow-bellied Prinia is olive above, with a grey head, whitish throat and buffy yellowish belly. It is common in all forms of wet grasslands including rank rice-fields, up to an elevation of 1300 metres, where its cheerful little 'seeleelak' call is a constant companion.

The small ZITTING CISTICOLA (W) (or Fantail-warbler) *Cisticola juncidis* (11 cm) occurs in similar open, damp habitats. It is a streaky brown warbler whose song consists of merely a 'zit' note at about one-second intervals, uttered in dipping song-flight.

FLYEATER
Gerygone sulphurea (9 cm) Plate 17g

The Flyeater is best known for its song, as the bird is generally hidden in the canopy. Especially common in mangroves, it occurs in a wide range of wooded habitats, including casuarinas along the beach, rubber, and sometimes secondary growth and forest. The song consists of repeated phrases of about three to five notes on a descending scale, wheezy and uttered as if with effort. Once learnt, this call is easily remembered. The rounded head and rather large bill are distinctive features.

Flycatchers

THE flycatchers form a large family of woodland insect-eaters, with some thirty-five species in West Malaysia, including some migrants from the north. They tend to be quiet and unobtrusive, and although some are quite colourful, few are easy to identify, especially the brown species in the genera *Rhinomyias* and *Muscicapa*. Many have songs, but these tend to be short, rather thin phrases and they sound rather alike. The greatest variety of species is found in the hills. Space can be given here only to a selection of representative species.

HILL BLUE FLYCATCHER
Cyornis banyumas (15 cm) Plate 20a

The males of this group of flycatchers have dark blue upper-parts, often more shining blue on the forehead, and bright rufous underparts. The females are usually browner. The Hill Blue Flycatcher is found locally in the hill forests, while the very similar MANGROVE BLUE FLYCATCHER *C. rufigastra* occurs in coastal areas. Lowland forests have the MALAYSIAN BLUE FLYCATCHER *C. turcosus*, in which the throat of the male is also blue; a rather uncommon bird, it is found only close to water.

LARGE NILTAVA (W)

Niltava grandis (22 cm) Plate 20b

The mountain forests have some all blue flycatchers. This is a large, rather thrush-like bird, quite common at Fraser's Hill and Cameron Highlands, in which the male appears almost black in a poor light. The female is rufous brown but with a blue patch in the neck. A rather fearless bird that is active close to ground level, it has a clear, rising whistle of three or four notes.

The VERDITER FLYCATCHER *Eumyias thalassina* (16 cm) is greenish-blue all over, and lives in the upper storey of hill forest up to about 1000 metres.

RUFOUS-CHESTED FLYCATCHER

Ficedula dumetoria (11 cm) Plate 20d

Another group of flycatchers has the upper-parts very dark grey to black. This species has a rufous breast, with a paler throat and belly, and white patches above the eye, in the wing, and on the sides to the base of the tail. It is a quiet bird of the understorey, mainly in the lower hills, quite tame and not difficult to watch, but it is not very common.

In the higher mountains, the charming, tame little SNOWY-BROWED FLYCATCHER *F. hyperythra* is dark bluish-grey above, with a short, thick white eye-brow in front of the eye.

GREY-HEADED FLYCATCHER

Culicicapa ceylonensis (13 cm) Plate 20c

The Grey-headed Flycatcher is common in the middle storey of the forest in the lowlands and hills. It has a warbler-like appearance and behaviour. The entire head region is grey, the upper-parts are olive and the under-parts yellow. Its simple four-note call, alternately low and high, 'silly billy', often reveals its presence.

PIED FANTAIL
Rhipidura javanica (18 cm)

As the name implies, the fantails have rather long tails which are frequently spread like a fan, exposing the white tips of the feathers. The Pied Fantail is a familiar bird of scrub, especially in damp places, and it is equally common in mangroves as in town gardens. The song is a short warble, loud and distinctive though difficult to describe.

In montane forests, it is replaced by the WHITE-THROATED FANTAIL *R. albicollis* which differs in being blackish-grey all over, except for a thin white stripe above the eye and a small white throat patch.

Pied Fantail

BLACK-NAPED MONARCH
Hypothymis azurea (14 cm) Plate 20e

This is the commonest member of a group of flycatchers known as the monarchs. Plate 20e shows it to be a very distinctive bird, unmistakable if seen clearly. The female is slightly duller, and lacks the black markings. It is found in the middle storey of forests in the lowlands and lower hills, where its 'wit-wit-wit' call is often heard.

MAROON-BREASTED FLYCATCHER
Philentoma velatum (20 cm) Plate 20f

This lovely bird is rather rare in lowland forests, frequenting the middle storey. It is a dark bird, and may appear black in a poor light.

1. (a) Brahminy Kite, p. 5. (b) Black-shouldered Kite, p. 5. (c) Crested Serpent-eagle, p. 6. (d) Japanese Sparrowhawk, p. 6.

2. (a) Lesser Whistling-duck, p. 7. (b) Crestless Fireback, p. 8. (c) Blue-breasted Quail, p. 9. (d) Crested Wood-partridge, p. 8. (e) Pheasant-tailed Jacana, p. 10.

3. (a) White-browed Crake, p. 10. (b) White-breasted Waterhen, p. 9. (c) Common Sandpiper, p. 11. (d) Golden Plover, p. 12. (e) Oriental Pratincole, p. 13.

4. (a) Little Green Pigeon, p. 15. (b) Pink-necked Pigeon, p. 16. (c) Little Cuckoo-dove, p. 16. (d) Mountain Imperial Pigeon, p. 16. (e) Peaceful Dove, p. 17. (f) Spotted Dove, p. 17.

5. (a) Jambu Fruit-dove, p. 17. (b) Green-winged Pigeon, p. 17.
(c) Blue-rumped Parrot, p. 18. (d) Long-tailed Parakeet, p. 18.
(e) Blue-crowned Hanging Parrot, p. 18.

6. (a) Plaintive Cuckoo, p. 19. (b) Indian Cuckoo, p. 20. (c) Raffles' Malkoha, p. 21. (d) Chestnut-bellied Malkoha, p. 22. (e) Greater Coucal, p. 22.

7. (a) Collared Scops Owl, p. 23. (b) Barn Owl, p. 24. (c) Buffy Fish-owl, p. 24 (d) Brown Hawk-owl, p. 24. (e) Javan Frogmouth, p. 25. (f) Collared Owlet, p. 23. (g) Malaysian Eared Nightjar, p. 25.

8. (a) Diard's Trogon, p. 29. (b) Red-naped Trogon, p. 29. (c) Scarlet-rumped Trogon, p. 29. (d) Orange-breasted Trogon, p. 30.

9. (a) Rufous-backed Kingfisher, p. 30. (b) Common Kingfisher, p. 30. (c) Collared Kingfisher, p. 31. (d) Rufous-collared Kingfisher, p. 31. (e) White-throated Kingfisher, p. 31. (f) Stork-billed Kingfisher, p. 32.

10. (a) Helmeted Hornbill, p. 37. (b) Great Hornbill, p. 34.
(c) Rhinoceros Hornbill, p. 34. (d) Wreathed Hornbill, p. 36.
(e) Wrinkled Hornbill, p. 36.

11. (a) Blue-throated Bee-eater, p. 32. (b) Red-bearded Bee-eater, p. 33. (c) Dollarbird, p. 33. (d) Red-crowned Barbet, p. 38. (e) Blue-eared Barbet, p. 38. (f) Gold-whiskered Barbet, p. 38. (g) Coppersmith Barbet, p. 39. (h) Brown Barbet, p. 39.

12. (a) Red-throated Barbet, p. 38. (b) Yellow-crowned Barbet, p. 39.
(c) Fire-tufted Barbet, p. 40. (d) Black-browed Barbet, p. 40.
(e) Golden-throated Barbet, p. 40.

13. (a) Brown-capped Woodpecker, p. 41. (b) Grey-and-Buff Woodpecker, p. 42. (c) Bay Woodpecker, p. 42. (d) Crimson-winged Woodpecker, p. 43. (e) Greater Yellownape, p. 43. (f) Common Goldenback, p. 44.

14. (a) Black-and-Yellow Broadbill, p. 45. (b) Banded Broadbill, p. 45. (c) Dusky Broadbill, p. 46. (d) Black-and-Red Broadbill, p. 45. (e) Long-tailed Broadbill, p. 46. (f) Green Broadbill, p. 46.

15. (a) Garnet Pitta, p. 47. (b) Blue-winged Pitta, p. 47. (c) Red-rumped Swallow, p. 48. (d) Common Iora, p. 55. (e) Scarlet Minivet, p. 52. (f) Blue-winged Leafbird, p. 55. (g) Fairy Bluebird, p. 56.

16. (a) Yellow-vented Bulbul, p. 53. (b) Straw-headed Bulbul, p. 53. (c) Black-crested Bulbul, p. 54. (d) Black-headed Bulbul, p. 53. (e) Scaly-breasted Bulbul, p. 54. (f) Red-eyed Bulbul, p. 54. (g) Yellow-bellied Bulbul, p. 54.

17. (a) White-rumped Shama, p. 57. (b) Chestnut-naped Forktail, p. 58. (c) Blue Rock-thrush, p. 58. (d) Dark-necked Tailorbird, p. 64. (e) Common Tailorbird, p. 65. (f) Yellow-bellied Prinia, p. 65. (g) Flyeater, p. 66.

18. (a) White-chested Babbler, p. 59. (b) Black-capped Babbler, p. 59. (c) Rufous-crowned Babbler, p. 60. (d) Large Wren-babbler, p. 60. (e) Black-throated Babbler, p. 61. (f) Chestnut-backed Scimitar-babbler, p. 62. (g) Fluffy-backed Tit-babbler, p. 61.

19. (a) Black Laughingthrush, p. 62. (b) Chestnut-capped Laughingthrush, p. 62. (c) Silver-eared Mesia, p. 63. (d) Cutia, p. 63. (e) Blue-winged Siva, p. 63. (f) White-browed Shrike-babbler, p. 63. (g) Brown Fulvetta, p. 64. (h) Malaysian Rail-babbler, p. 64.

20. (a) Hill Blue Flycatcher, p. 66. (b) Large Niltava, p. 67. (c) Grey-headed Flycatcher, p. 67. (d) Rufous-chested Flycatcher, p. 67. (e) Black-naped Monarch, p. 68. (f) Maroon-breasted Flycatcher, p. 68. (g) Mangrove Whistler, p. 70.

21. (a) Velvet-fronted Nuthatch, p. 71. (b) Crimson-breasted Flowerpecker, p. 72. (c) Yellow-breasted Flowerpecker, p. 72. (d) Scarlet-backed Flowerpecker, p. 71. (e) Buff-bellied Flowerpecker, p. 72. (f) Orange-bellied Flowerpecker, p. 71.

22. (a) Brown-throated Sunbird, p. 73. (b) Copper-throated Sunbird, p. 73. (c) Olive-backed Sunbird, p. 73. (d) Black-throated Sunbird, p. 74. (e) Purple-naped Sunbird, p. 73. (f) Little Spiderhunter, p.74. (g) Oriental White-eye, p. 74.

23. (a) Eurasian Tree Sparrow, p. 75. (b) Plain-backed Sparrow, p. 75. (c) Baya Weaver, p. 75. (d) Chestnut Munia, p. 76. (e) Scaly-breasted Munia, p. 76. (f) Pin-tailed Parrotfinch, p. 76.

24. (a) Hill Myna, p. 77. (b) Common Myna, p. 77. (c) Asian Glossy Starling, p. 77. (d) Black-naped Oriole, p. 78. (e) Dark-throated Oriole, p. 78. (f) Green Magpie, p. 81.

Not being very active, it is readily overlooked, unless its whirring wings in flight or its rather fluty, short monotone whistle reveals its presence. This whistle is sometimes succeeded by a single or double harsh 'chak', slightly recalling the whiplash notes of a whistler (see p. 70). The female is duller and lacks the maroon breast.

ASIAN PARADISE-FLYCATCHER

Terpsiphone paradisi (20 cm, plus elongated tail plumes in the male)

The remarkable male Asian Paradise-flycatcher is a white bird, with a black head and primaries, and two white tail plumes up to 23 centimetres long. It is quite common in lowland forests, and of course it is unmistakable. However, the female is not white at all, being deep rufous, blacker around the head, and paler below, and lacking the elongated tail. Some males also occur in a rufous phase which is similar to the female. The calls are similar to those of the Black-naped Monarch.

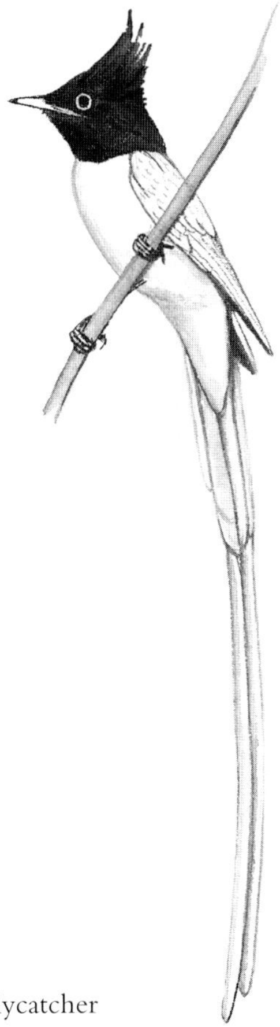

Asian Paradise Flycatcher

MANGROVE WHISTLER
Pachycephala grisola (15 cm) Plate 20g

The Mangrove Whistler is a representative of a large Australian family. It is rather nondescript, greyish-brown, paler below, with a noticeably rounded head. The most distinctive feature is its varied song, which typically consists of a series of loud whistled notes on one pitch followed by a single or double whiplash note. (This should not be confused with the 'weeeee-tu' of the Common Iora, see p. 55.) It has, however, a wide range of phrases, including warbled versions, but the explosive notes usually give it away. It is not confined to mangroves, being equally found in a variety of lowland wooded habitats, including the thinner forests of sandy or peaty soils, and old rubber plantations.

Tits and Nuthatches

SULTAN TIT (W)
Melanochlora sultanea (20 cm)

This bird is unmistakable with its black plumage and brilliant yellow crown, crest, and belly. The black has a slightly greenish tinge, especially in the female. It is a common bird of forests and forest margins in the foothills, more especially in the north.

Sultan Tit

VELVET-FRONTED NUTHATCH

Sitta frontalis (13 cm) Plate 21a

Small, compact, short-tailed birds, nuthatches clamber about the trunks and branches of trees in woodland, and have the unique ability to climb vertically down the trunk, facing down. This is a bird of lowland forests, which has all whitish under-parts and a red bill.

Above 1000 metres, it is replaced by the BLUE NUTHATCH (W) *S. azurea*, which differs in having bluish black upper-parts and belly, only the throat and breast being white; the bill is light blue.

Flowerpeckers

FLOWERPECKERS are tiny, active little birds, constantly flitting from one flowering shrub to another in swift but erratic flight. The males especially are very colourful, but they do not have the glossy metallic sheen of the sunbirds, nor the latter's delicate, curved bills. They are particularly partial to the epiphytes growing on forest trees. There are fourteen species in our region. Only the more colourful and obtrusive species are described here.

SCARLET-BACKED FLOWERPECKER

Dicaeum cruentatum (9 cm) Plate 21d

The male of this common species of gardens and wooded areas is distinctive with its black body, red crown, back, and rump, and whitish streak down the centre of the under-parts. The female is greyer, with the red confined to the back.

ORANGE-BELLIED FLOWERPECKER

Dicaeum trigonostigma (9 cm) Plate 21f

Another distinctive species is the Orange-bellied Flowerpecker, with its orange-yellow back, lower breast and belly, pale grey throat, and dark blue-grey crown, mantle, and wings. The female is slightly less colourful (see Plate 21f). It is a common forest and woodland bird of the lowlands and lower mountains.

BUFF-BELLIED FLOWERPECKER (W)
Dicaeum ignipectus (9 cm) Plate 21e

The Buff-bellied (or Fire-breasted) Flowerpecker is a montane flowerpecker, with blue-green upper-parts and, in the male, a scarlet patch on the breast and a black line down the centre of the belly.

CRIMSON-BREASTED FLOWERPECKER
Prionochilus percussus (10 cm) Plate 21b

The Crimson-breasted Flowerpecker is a common bird of the middle storey of primary and secondary forest in the lowlands and hills. The male recalls the previous species with its crimson breast patch, and both sexes have a red patch on the crown.

YELLOW-BREASTED FLOWERPECKER
Prionochilus maculatus (9 cm) Plate 21c

The Yellow-breasted Flowerpecker is one of the plainer species, olive-green above and yellow below with olive streaks. It has a small orange patch on the crown. It is quite common in lowland forests.

A much rarer but especially beautiful bird is the SCARLET-BREASTED FLOWERPECKER *P. thoracicus*, in which the male has black fore-parts with a prominent scarlet patch on the breast and a smaller one on the crown. The back is olive and the belly is yellow.

Sunbirds and Spiderhunters

SUNBIRDS might be considered as the oriental equivalent of the Neotropical hummingbirds, being tiny, active little nectar feeders with slender, curved bills. They feed on both nectar and insects, and are always busy about flowering shrubs though they do not have the capacity of prolonged hovering. The males have brilliant glossy plumage, though in the field this often looks black. Females, young, and males in moult have dull plumage and they are very difficult to identify. The region has eleven sunbirds and seven of the very long-billed spiderhunters.

BROWN-THROATED SUNBIRD
Anthreptes malacensis (14 cm) Plate 22a

A common and widespread sunbird of open country, this species especially favours coastal coconut groves. The main features in the male are the brown throat and yellow belly. The upper-parts are metallic green, grading to purple on the rump and wing coverts. By contrast, the female is olive-brown above and yellow below, a broad description that fits most female sunbirds.

OLIVE-BACKED SUNBIRD
Nectarinia jugularis (11 cm) Plate 22c

The male of this species is distinguished by its olive upper-parts. The throat is a metallic near-black, and the belly is bright yellow. All the under-parts of the female are yellow. Useful characters are the rather longer and thinner bill than in the Brown-throated Sunbird and, especially, the white lateral feathers on the dark tail. It is very common in open habits in the lowlands.

COPPER-THROATED SUNBIRD
Nectarinia calcostetha (14 cm) Plate 22b

This species occurs mainly in the coastal regions, especially mangroves. The male is almost wholly dark, with a variety of metallic colours, and with orange pectoral tufts beneath the wings, though these are often concealed. The tail is slightly graduated, and in the female it is tipped white.

PURPLE-NAPED SUNBIRD
Hypogramma hypogrammicum (15 cm) Plate 22e

This large sunbird is quite distinctive with its streaked under-parts and, in the male, the metallic purple bar across the nape. There is also purple on the rump though this is less conspicuous. It is quite common in scrubby growth in the forest edge, often near rivers or swamps.

BLACK-THROATED SUNBIRD (W)
Aethopyga saturata (13 cm) Plate 22d

The common sunbird of the forests and gardens in the hill stations is the Black-throated Sunbird, one of a genus in which the males have elongated central tail feathers. The male has dark metallic upper-parts and throat, and a yellow lower breast.

The CRIMSON SUNBIRD *A. siparaja* (11 cm) is a related species of the coastal regions, and the male is a particularly beautiful bird, being crimson red with a glossy dark forecrown, moustachial stripe and tail, and a yellow back. The belly is dark grey. Inland in the foothills, it is replaced by the very locally distributed SCARLET SUNBIRD *A. temminckii*, which is almost identical except that the long tail is red, and the belly is a lighter grey.

LITTLE SPIDERHUNTER
Arachnothera longirostra (16 cm) Plate 22f

Spiderhunters are similar to sunbirds but have extremely long, decurved bills and mainly green plumage in both sexes. This is perhaps the commonest and also the most readily recognizable species, because of its yellow belly and greyish-white throat. A bird living mainly in the forest edge, it especially favours wild bananas and ginger. The flight is direct and swift. The call is a single note that can be repeated with great monotony.

The other spiderhunters have very similar plumage, and it is not easy to get good views, so they are difficult birds for the novice. However, the STREAKED SPIDERHUNTER (W) *A. magna* (19 cm), which is the common species in the mountains, is readily identified by the heavy black streaks on the body and the orange legs.

White-eyes

ORIENTAL WHITE-EYE
Zosterops palpebrosus (11 cm) Plate 22g

The white-eyes are active birds of the canopy that behave rather like sunbirds or flowerpeckers. They usually occur in loose parties that maintain contact by calling incessantly, with cheeping notes rather like baby chickens. They are mainly olive in colour, and are so named from the prominent white ring around the eye. There are two species in our region, but they are not easy birds for the amateur to distinguish. The Oriental White-eye is quite common in the coastal mangroves and adjacent woodlands, and is replaced inland and in the hills by the slightly darker EVERETT'S WHITE-EYE *Z. everetti*.

Sparrows, Weavers, and Munias

THIS is the family of sparrows and seed-eaters that plunder the rice-fields, the so-called paddy or rice birds. They are small, occasionally colourful, rounded, gregarious birds with short, thick bills. The Malay name for the munias is *Pipit*, which should not be confused with the English name for birds in the genus *Anthus* (see p. 49).

PLAIN-BACKED SPARROW (or Pegu Sparrow) (W)
Passer flaveolus (15 cm) Plate 23b

This attractive, cleanly coloured sparrow is extending its range down the coastal regions of the peninsula, presently south to Selangor and Kuantan. It lives in the open rice-fields and coastal plantations, and is only incidentally associated with Man.

EURASIAN TREE SPARROW
Passer montanus (14 cm) Plate 23a

This is the drab sparrow, with dark chocolate-brown cap and black bib and face spot, that follows Man to all his habitations, and is common everywhere in towns and villages.

BAYA WEAVER (W)
Ploceus philippinus (14 cm) Plate 23c

Weavers are the birds that build the colonies of elaborate woven suspended nests in palm fronds and isolated bushes in the rice-fields. The birds are noisy and conspicuous in the breeding season, with cheerful chatters and wheezes, and the males acquire the plumage shown in

Plate 23c. Outside the breeding season, the males resemble the dowdy females, which lack the yellow crown and black mask.

CHESTNUT MUNIA
Lonchura malacca (11 cm) Plate 23d

Two of the common munias of open country have chestnut bodies. This species is a rich chestnut, with black fore-parts. It often associates with the WHITE-HEADED MUNIA *L. maja*, which is chestnut-brown with white fore-parts. Both occur in 'cheeping' flocks in rice-fields, grasslands, and secondary growth.

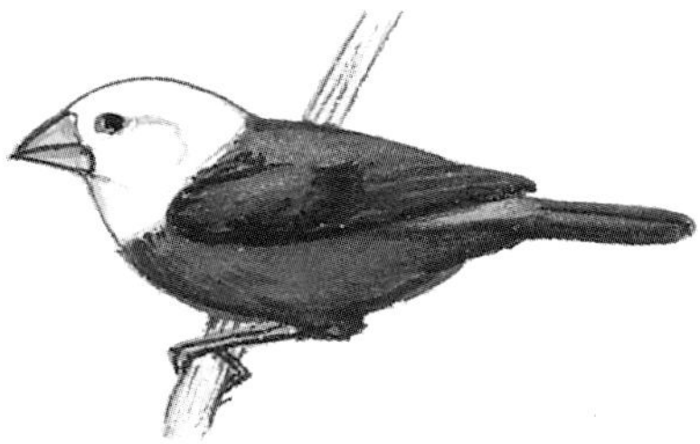

White-headed Munia

SCALY-BREASTED MUNIA (W)
Lonchura punctulata (11 cm) Plate 23e

Often associating with other munias, this bird has less striking plumage, but still distinctive with its scaly under-parts and reddish front of the head.

A less colourful bird found in both cultivation and forest clearings is the WHITE-RUMPED MUNIA (W) *L. striata*, which is blackish-brown with a prominent white rump and a white belly; its tail is pointed. The similar WHITE-BELLIED MUNIA *L. leucogastra* differs in having a dark rump and rounded tail, and is a rather uncommon bird of forest margins.

PIN-TAILED PARROTFINCH
Erythrura prasina (13 cm) Plate 23f

This extremely colourful little bird is much more secretive, living in bamboos and light forest usually in hilly terrain, never venturing far from this habitat. It lives in parties and seems to be nomadic and erratic in its movements, following the seeding of the large bamboos.

Mynas and Starlings

HILL MYNA
Gracula religiosa (30 cm) Plate 24a

Also known as Grackle, or Talking Myna, this bird is famous for its seeming ability to talk, although pet birds are only imitating the sounds their owners have taught them. A large, bluish-black bird with heavy orange bill and yellow head-lappets, there is a white patch in the rounded wings, prominent in flight. Its most typical call is a piercing 'ti-ong'. It is not uncommon in twos and threes in lowland forest, where it keeps to the upper canopy.

COMMON MYNA
Acridotheres tristis (25 cm) Plate 24b

The Common Myna has spread south since the 1930s and is now common everywhere in open country. It has a black head grading to a brown body with a white belly, a white patch in the wings, white tip to the tail, and yellow bill and cheek patch.

At about the same time as the Common Myna was spreading south, the WHITE-VENTED MYNA (W) *A. javanicus* was introduced into Singapore from Java, and is now common in the south and spreading elsewhere. This is black and greyish-black, with a small crest, and with a similar distribution of white on the belly, wings, and tail; it also has a yellow bill, and a yellow eye. These are the two common species, but there are other mynas, both feral and native (or resulting perhaps from earlier range expansions), and their taxonomy is somewhat confusing.

ASIAN GLOSSY STARLING
Aplonis panayensis (20 cm) Plate 24c

The Asian Glossy Starling is a common bird in open country and towns, especially in coastal regions, where large numbers may congregate to roost in shrill parties in selected groves of coconut palms. It is a sleek, glossy green bird with red eyes, although it looks black in flight. Young birds are whitish below with dark streaks.

The PURPLE-BACKED STARLING *Sturnus sturninus* (19 cm) is a common winter visitor, in small flocks mainly to the open coastal lowlands, often seen together with Asian Glossy Starlings. It is black above and white below, with a grey head, whitish rump, and white wing bar, and the female is browner.

Orioles

BLACK-NAPED ORIOLE (W)
Oriolus chinensis (27 cm) Plate 24d

Most orioles have golden yellow plumage. This species has a black stripe through the face and round the back of the neck, and also black primaries and central tail feathers. Its fluty 'ori-ol-lo' notes delight the residents of towns and suburbs, where it lives in the trees in avenues and gardens. However, in contrast, its call note is very harsh. It is a rather recent arrival here, while another form is a winter migrant in the northern states.

DARK-THROATED ORIOLE
Oriolus xanthonotus (19 cm) Plate 24e

This smaller oriole is a common bird of forests in the lowlands and lower hills. As it remains mostly in the canopy, it is difficult to see, but hardly a day passes in the forest without hearing its fluty whistle at least once.

A lovely forest oriole of the mountains is the BLACK-AND-CRIMSON ORIOLE *O. cruentatus* (22 cm), which is black with a crimson lower breast. However, being a canopy bird, it is difficult to observe and it appears all black in a poor light. The call is a nasal mew.

Drongos

THESE are black or grey birds with rather long and strikingly forked tails. They are notoriously aggressive, driving off birds twice their size, even eagles and hornbills. They have a wide range of call notes, sometimes mimicking other species, and the calls do not help in identification.

GREATER RACKET-TAILED DRONGO
Dicrurus paradiseus (33 cm, plus up to 30 cm of tail rackets)

This common drongo has enormously lengthened outer tail feathers, each bearing a spoon-shaped racket at the tip. The rackets are often twisted. However, in many individuals, these elongated feathers are missing, either through abnormal development or when the bird is in moult. At such times, it is not so easy to separate from the other black drongos. The similar LESSER RACKET-TAILED DRONGO (W) *D. remifer* replaces it in the mountains, and differs in having flat, elongate rackets.

A much smaller bird is the BRONZED DRONGO *D. aeneus* (24 cm), which is slender and glossy black with a deeply forked tail, without rackets. It is a common forest bird of the lowlands and lower hills.

The ASHY DRONGO *D. leucophaeus* (28 cm) is grey with a moderately forked tail. The resident form in the open plains of the west coast, south nearly to Malacca, is almost black on the upper-parts, but paler below. In winter, however, a few migrants reach the region which are much paler grey.

Greater Racquet-tailed Drongo

Crows

THE omnivorous black crows are familiar to most people, but the family includes a range of quite interesting and sometimes colourful birds.

LARGE-BILLED CROW (W)
Corvus macrorhynchos (51 cm)

This is the common crow of open country, a heavy black bird, usually seen in twos and threes, uttering guttural croaking notes. However, it is being replaced in the towns and plains of the west coast by the HOUSE CROW (W) *C. splendens* (44 cm). This very noisy and familiar black and grey crow is a recent arrival in the peninsula, both through deliberate introduction and also on board ship.

Large-billed Crow

CRESTED JAY
Platylophus galericulatus (33 cm)

This very distinctive looking bird is not often seen, as it stays deep in the forest, but it is quite common. Its call often reveals its presence, and consists of a harsh rattle, readily recognized once known. It is a frequent participant in the mixed feeding flocks that are such a common feature of tropical forests.

Crested Jay

More readily seen is the BLACK MAGPIE *Platysmurus leucopterus* (40 cm), which prefers forest edges and more open woodland, including neglected rubber estates. This is a black bird with a moderately long tail, a small crest, and a white patch on the wing. It has a variety of peculiar calls, including bell-like notes uttered singly or in short ascending runs, as well as more discordant squawks.

GREEN MAGPIE

Cissa chinensis (38 cm) Plate 24f

One would expect these beautiful birds to be easy to see, but they are quite shy birds of montane forests and scrub, often only detected when their loud calls give them away. Generally an observer will get only short glimpses as single birds of a party fly from thicket to thicket.

Appendices

APPENDIX 1
Birds in East Malaysia

THE birds marked (W) in the text do not occur in East Malaysia (or, in a few cases, occur on the island of Borneo but are very rare). For most of the remaining birds, the descriptions apply equally to the eastern states. However, bird-watchers in this region should be aware of the presence of the following species, which are endemic to the island of Borneo. Many of these are confined to montane areas.

Birds endemic to Borneo

Kinabalu Serpent-eagle	*Spilornis kinabaluensis*
White-fronted Falconet	*Microhierax latifrons*
Red-breasted Partridge	*Arborophila hyperythra*
Crimson-headed Partridge	*Haematortyx sanguiniceps*
Bulwer's Pheasant	*Lophura bulweri*
Bornean Peacock-pheasant	*Polyplectron schleiermacheri*
Mantanani Scops Owl	*Otus mantanensis*
Dulit Frogmouth	*Batrachostomus harterti*
Whitehead's Trogon	*Harpactes whiteheadi*
Mountain Barbet	*Megalaima monticola*
Black-throated Barbet	*Megalaima eximia*
Golden-naped Barbet	*Megalaima pulcherrima*
Hose's Broadbill	*Calyptomena hosii*
Whitehead's Broadbill	*Calyptomena whiteheadi*
Blue-banded Pitta	*Pitta arquata*
Blue-headed Pitta	*Pitta baudii*
Fruithunter	*Chlamydochaera jefferyi*
Bornean Bristlehead	*Pityriasis gymnocephala*
Everett's Thrush	*Zoothera everetti*
Black-browed Babbler	*Trichastoma perspicillatum*
Bornean Wren-babbler	*Ptilocichla leucogrammica*
Black-throated Wren-babbler	*Napothera atrigularis*
Mountain Wren-babbler	*Napothera crassa*
Bald-headed Laughing-thrush	*Garrulax calvus*

Chestnut-crested Yuhina	*Yuhina everetti*
Bornean Stubtail	*Urosphena whiteheadi*
Friendly Bush-warbler	*Bradypterus accentor*
Bornean Blue Flycatcher	*Cyornis superbus*
Bornean Whistler	*Pachycephala hypoxantha*
Yellow-rumped Flowerpecker	*Prionochilus xanthopygius*
Black-sided Flowerpecker	*Dicaeum monticolum*
Whitehead's Spiderhunter	*Arachnothera juliae*
Pigmy Darkeye	*Oculicincta squamifrons*
Blackeye	*Chlorocharis emiliae*
Dusky Munia	*Lonchura fuscans*
Black Oriole	*Oriolus hosii*

Descriptions of these will be found in *The Birds of Borneo* by B. E. Smythies (3rd edn., 1981), published by The Sabah Society and the Malayan Nature Society. This book has also been reproduced as a pocket guide (the illustrations and brief notes only), published by The Sabah Society and World Wildlife Fund Malaysia in 1984.

APPENDIX 2
Malay Names of Birds

Brown Booby	*Dendang Laut*
Cattle Egret	*Bangau Kendi*
Great Egret	*Bangau Besar*
Reef Egret	*Bangau Batu*
Chinese Pond-heron	*Pucong Cina*
Grey Heron	*Pucong Seriap*
Purple Heron	*Pucong Serandau*
Little Heron	*Pucong Keladi*
Cinnamon Bittern	*Pucong Bendang*
Lesser Adjutant	*Burung Botak Kecil*
Milky Stork	*Burung Upeh*
Brahminy Kite	*Lang Merah*
Black-shouldered Kite	*Lang Bahu Hitam*
White-bellied Sea-eagle	*Lang Siput*
Crested Serpent-eagle	*Lang Berjambul/Lang Kuik*
Black Eagle	*Lang Hitam*
Changeable Hawk-eagle	*Lang Hindek*
Japanese Sparrowhawk	*Lang Sewah*
Oriental Honey-buzzard	*Lang Lebah*
Black-thighed Falconet	*Rajawali*
Lesser Whistling-duck	*Belibis*
Great Argus	*Kuang Raya*
Crested Argus	*Kuang Raya Gunung*
Crestless Fireback	*Merah Mata*
Crested Fireback	*Ayam Pegar*
Crested Wood-partridge	*Burung Siul*
Long-billed Partridge	*Burung Selanting*
Red Junglefowl	*Ayam Hutan*
Blue-breasted Quail	*Pikau*
White-breasted Waterhen	*Ruak-ruak*
Moorhen	*Tiong Air*
White-browed Crake	*Sintar Dahi Putih*
Pheasant-tailed Jacana	*Burung Teratai Besar*
Masked Finfoot	*Pedendang*
Golden Plover	*Rapang Kerinyut*
Stints	*Kedidi kecil*

Greenshank	*Kedidi Kaki Hijau*
Common Sandpiper	*Kedidi Pasir*
Wood Sandpiper	*Kedidi Kayu*
Curlew	*Kedidi Kendi*
Whimbrel	*Kedidi Pisau Raut*
Oriental Pratincole	*Lelayang Padang*
Black-naped Tern	*Camar Topi Hitam*
Little Tern	*Camar Kecil*
Great Crested Tern	*Camar Besar Berjambul*
Lesser Crested Tern	*Camar Kecil Berjambul*
White-winged Tern	*Camar Kepak Putih*
Whiskered Tern	*Camar Batu Berumbai*
Little Green Pigeon	*Punai Daun*
Thick-billed Pigeon	*Punai Lengguak*
Pink-necked Pigeon	*Punai Gading*
Cinnamon-headed Pigeon	*Punai Bakau*
Mountain Imperial Pigeon	*Pergam Gunung*
Green Imperial Pigeon	*Pergam Besar*
Little Cuckoo-dove	*Tekukur Api*
Spotted Dove	*Tekukur/Merbok Balam*
Peaceful Dove	*Merbok Aman*
Jambu Fruit-dove	*Punai Jambu*
Green-winged Pigeon	*Punai Tanah*
Long-tailed Parakeet	*Bayan Nuri*
Blue-rumped Parrot	*Bayan Puting*
Blue-crowned Hanging Parrot	*Serindit*
Plaintive Cuckoo	*Sewah Mati Anak*
Brush Cuckoo	*Sewah Gila*
Banded Bay Cuckoo	*Sewah Takuweh*
Indian Cuckoo	*Sewah India*
Large Hawk-cuckoo	*Sewah Tekukur Besar*
Violet Cuckoo	*Sewah Rembah*
Drongo Cuckoo	*Sewah Sawai*
Common Koel	*Sewah Tahu*
Raffles' Malkoha	*Cenok Kerak*
Chestnut-breasted Malkoha	*Cenok Birah*
Red-billed Malkoha	*Cenok Api*
Chestnut-bellied Malkoha	*Cenok Kecil*
Black-bellied Malkoha	*Cenok Perut Hitam*
Green-billed Malkoha	*Cenok Kera*

Greater Coucal	*Butbut Carik Anak*
Lesser Coucal	*Butbut Kecil*
Collared Scops Owl	*Burung Hantu Reben*
Mountain Scops Owl	*Burung Hantu Gunung*
Collared Owlet	*Burung Hantu Kecil*
Brown Hawk-owl	*Burung Hantu Bertemak*
Barn Owl	*Jampok Kubur*
Buffy Fish-owl	*Burung Hantu Kuning*
Barred Eagle-owl	*Burung Hantu Bubu*
Malaysian Eared Nightjar	*Teptibau/Tukang Malaysia*
Great Eared Nightjar	*Tukang Telinga Besar*
Large-tailed Nightjar	*Tukang Kubur*
Javan Frogmouth	*Segan Jawa*
House Swift	*Layang-layang Rumah*
Palm Swift	*Layang-layang Asia*
White-bellied Swiftlet	*Layang-layang Perut Putih*
Edible-nest Swiftlet	*Layang-layang Gua*
Black-nest Swiftlet	*Layang-layang Padi*
Silver-rumped Swift	*Layang-layang Kecil*
Brown Needletail	*Layang-layang Besar*
Whiskered Treeswift	*Layang-layang Jambul Kecil*
Grey-rumped Treeswift	*Layang-layang Jambul Kelabu*
Diard's Trogon	*Kasumba Bukit*
Red-naped Trogon	*Kasumba Batang*
Scarlet-rumped Trogon	*Kasumba Puteri*
Orange-breasted Trogon	*Kasumba Harimau*
Rufous-backed (Dwarf) Kingfisher	*Binti-binti (or Pekaka Api)*
Black-backed (Dwarf) Kingfisher	*Binti-binti (or Pekaka Rimba)*
Common Kingfisher	*Pekaka Citcit Kecil*
Blue-eared Kingfisher	*Pekaka Bintik-bintik*
Rufous-collared Kingfisher	*Pekaka Rimba Besar*
Banded Kingfisher	*Pekaka Riang Rimba*
Collared Kingfisher	*Pekaka Sungai*
White-throated Kingfisher	*Pekaka Belukar*
Black-capped Kingfisher	*Pekaka Kopiah Hitam*
Stork-billed Kingfisher	*Pekaka Paruh Besar*
Blue-throated Bee-eater	*Berek-berek Tadah Hujan*
Blue-tailed Bee-eater	*Berek-berek Carik Dada*
Chestnut-headed Bee-eater	*Berek-berek Senja*
Red-bearded Bee-eater	*Berek-berek Janggut Merah*

Dollarbird	*Tiong Batu*
Rhinoceros Hornbill	*Enggang Badak*
Great Hornbill	*Enggang Papan*
Pied Hornbill	*Enggang Kelingking*
Black Hornbill	*Enggang Gatal Birah*
Bushy-crested Hornbill	*Enggang Belukar*
White-crowned Hornbill	*Enggang Jambul Putih*
Wreathed Hornbill	*Enggang Gunung*
Wrinkled Hornbill	*Enggang Berkedut*
Helmeted Hornbill	*Enggang Tebang Mertua*
Red-crowned Barbet	*Takor Kepala Merah*
Blue-eared Barbet	*Takor Akar*
Gold-whiskered Barbet	*Takor Jambang Emas*
Red-throated Barbet	*Takor Raya*
Yellow-crowned Barbet	*Takor Mahkota Kuning*
Coppersmith Barbet	*Takor Tembaga*
Lineated Barbet	*Takor Kukup*
Brown Barbet	*Takor Dahan*
Black-browed Barbet	*Takor Bukit*
Fire-tufted Barbet	*Takor Api*
Golden-throated Barbet	*Takor Leher Emas*
Rufous Piculet	*Belatok Kecil*
Speckled Piculet	*Belatok Belang*
Brown-capped Woodpecker	*Belatok Belacan Kecil*
Grey-and-buff Woodpecker	*Belatok Punggoh*
Buff-rumped Woodpecker	*Belatok Awan*
Buff-necked Woodpecker	*Belatok Tuki-tuki*
Bay Woodpecker	*Belatok Bulan*
Maroon Woodpecker	*Belatok Punggor*
Rufous Woodpecker	*Belatok Biji Nangka*
Crimson-winged Woodpecker	*Belatok Mas*
Banded Woodpecker	*Belatok Merah*
Checker-throated Woodpecker	*Belatok Ranting*
Greater Yellownape	*Belatok Besar Tengkok Kuning*
Lesser Yellownape	*Belatok Kecil Tengkok Kuning*
Common Goldenback	*Belatok Pinang Muda*
Orange-backed Woodpecker	*Belatok Ranum*
White-bellied Woodpecker	*Belatok Gajah*
Great Slaty Woodpecker	*Belatok Berjalur*
Black-and-yellow Broadbill	*Takau Hitam Kuning*

Banded Broadbill	*Takau Rimba*
Black-and-red Broadbill	*Takau Rakit*
Dusky Broadbill	*Takau Rimba Hujan*
Green Broadbill	*Takau Selawit*
Long-tailed Broadbill	*Takau Injap*
Silver-breasted Broadbill	*Takau Tanda Hujan*
Garnet Pitta	*Pacat Kepala Merah*
Banded Pitta	*Pacat Bukit*
Blue-winged Pitta	*Pacat Sayap Biru*
Mangrove Pitta	*Pacat Bakau*
Pacific Swallow	*Sualo Batu*
Barn Swallow	*Sualo Api*
Red-rumped Swallow	*Sualo Gua*
Grey Wagtail	*Pipit Batu*
Yellow Wagtail	*Pipit Kuning*
Richard's Pipit	*Pipit Tanah*
Bar-winged Flycatcher-shrike	*Rembah Bukit*
Black-winged Flycatcher-shrike	*Rembah Batu*
Pied Triller	*Burung Sewah Kapas*
Malaysian Cuckoo-shrike	*Sewah Besar*
Lesser Cuckoo-shrike	*Sewah Kecil*
Bar-bellied Cuckoo-shrike	*Sewah Rimba*
Scarlet Minivet	*Burung Mas Belukar*
Grey-chinned Minivet	*Burung Mas Dagu Kelabu*
Fiery Minivet	*Burung Mas Tulin*
Ashy Minivet	*Burung Mas Padang*
Yellow-vented Bulbul	*Merbah Kapur*
Red-whiskered Bulbul	*Merbah Telinga Merah*
Straw-headed Bulbul	*Barau-barau*
Black-headed Bulbul	*Merbah Siam*
Black-crested Bulbul	*Merbah Jambul Hitam*
Scaly-breasted Bulbul	*Merbah Bersisik*
Grey-bellied Bulbul	*Merbah Kelabu*
Red-eyed Bulbul	*Merbah Mata Merah*
Yellow-bellied Bulbul	*Merbah Perut Kuning*
Common Iora	*Burung Kunyit Kacat*
Green Iora	*Burung Kunyit Bakau*
Blue-winged Leafbird	*Burung Daun Sayap Biru*
Greater Green Leafbird	*Burung Daun Besar*
Lesser Green Leafbird	*Burung Daun Kecil*

Orange-bellied Leafbird	*Burung Daun Perut Oren*
Fairy Bluebird	*Burung Dendang Gajah*
Long-tailed Shrike	*Tirjup Ekor Panjang*
Brown Shrike	*Tirjup Tanah*
White-rumped Shama	*Murai Rimba*
Rufous-tailed Shama	*Murai Rimba Ekor Kuning*
Magpie Robin	*Murai Kampung*
Chestnut-naped Forktail	*Murai Cegar*
White-crowned Forktail	*Murai Cegar Belukar*
Blue Rock-thrush	*Murai Batu Tarung*
Blue Whistling Thrush	*Murai Batu Siul*
Malayan Whistling Thrush	*Murai Tiong Batu*
Black-capped Babbler	*Burung Rimba Kopiah Hitam*
White-chested Babbler	*Burung Rimba Telunjuk*
Abbott's Babbler	*Burung Rimba Riang*
Rufous-crowned Babbler	*Burung Rimba Tua Besar*
Scaly-crowned Babbler	*Burung Rimba Tua Kecil*
Sooty-capped Babbler	*Burung Rimba Tinjau Belukar*
Large Wren-babbler	*Burung Rimba Besar Hujan*
Streaked Wren-babbler	*Burung Rimba Tunggul*
Black-throated Babbler	*Burung Rimba Bertam*
Chestnut-winged Babbler	*Burung Rimba Merbah Sampah*
Chestnut-rumped Babbler	*Burung Rimba Rembah Besar*
Golden Babbler	*Burung Rimba Mas*
Fluffy-backed Tit-babbler	*Burung Rimba Pongpong*
Striped Tit-babbler	*Burung Rimba Berjalur*
Chestnut-backed Scimitar-babbler	*Burung Rimba Hulu*
Black Laughingthrush	*Burung Rimba Hitam*
Chestnut-capped Laughingthrush	*Burung Rimba Genting*
Silver-eared Mesia	*Burung Rimba Telinga Perak*
Cutia	*Burung Rimba Kutia*
Blue-winged Siva	*Burung Rimba Sayap Biru*
White-browed Shrike-babbler	*Burung Rimba Cekup Belalang*
Long-tailed Sibia	*Sibia Ekor Panjang*
Brown Fulvetta	*Burung Rimba Murai Coklat*
Malaysian Rail-babbler	*Burung Rimba Malaysia*
Dark-necked Tailorbird	*Perenjak Belukar*
Common Tailorbird	*Perenjak Pisang*
Rufous-tailed Tailorbird	*Perenjak Rimba*
Olive-backed Tailorbird	*Perenjak Belukar*

Mountain Tailorbird	*Perenjak Gunung*
Yellow-bellied Prinia	*Perenjak Padi*
Zitting Cisticola	*Cekup Layang*
Flyeater	*Cekup Perepat*
Hill Blue Flycatcher	*Burung Sambar Bukit*
Mangrove Blue Flycatcher	*Burung Sambar Biru Bakau*
Malaysian Blue Flycatcher	*Burung Sambar Biru Malaysia*
Large Niltava	*Burung Sambar Kumbang Padi*
Verditer Flycatcher	*Burung Sambar Ranting*
Rufous-chested Flycatcher	*Burung Sambar Dada Oren*
Snowy-browed Flycatcher	*Burung Sambar Kudong*
Grey-headed Flycatcher	*Burung Sambar Pacat*
Pied Fantail	*Burung Sambar Murai Gila*
White-throated Fantail	*Burung Sambar Gila Gunung*
Black-naped Monarch	*Burung Sambar Uban Hitam*
Maroon-breasted Flycatcher	*Burung Sambar Ungu*
Asian Paradise-flycatcher	*Burung Sambar Ekor Panjang*
Mangrove Whistler	*Burung Sambar Siul Belukar*
Sultan Tit	*Burung Serai Sultan*
Velvet-fronted Nuthatch	*Burung Patuk Baldu*
Blue Nuthatch	*Burung Patuk Gunung*
Crimson-breasted Flowerpecker	*Sepah Puteri Pelangi*
Yellow-breasted Flowerpecker	*Sepah Puteri Raja*
Scarlet-breasted Flowerpecker	*Sepah Puteri*
Scarlet-backed Flowerpecker	*Sepah Puteri Merah*
Orange-bellied Flowerpecker	*Sepah Puteri Bukit*
Buff-breasted Flowerpecker	*Sepah Puteri Sulong*
Brown-throated Sunbird	*Kelicap Mayang Kelapa*
Olive-backed Sunbird	*Kelicap Bukit*
Copper-throated Sunbird	*Kelicap Bakau*
Purple-naped Sunbird	*Kelicap Rimba*
Black-throated Sunbird	*Kelicap Gunung*
Crimson Sunbird	*Kelicap Sepah Raja*
Scarlet Sunbird	*Kelicap Merah*
Little Spiderhunter	*Kelicap Jantung*
Streaked Spiderhunter	*Kelicap Jantung Gunung*
Oriental White-eye	*Mata Putih Timur*
Everett's White-eye	*Mata Putih Belukar*
Plain-Backed Sparrow	*Ciak Arnab*
Eurasian Tree Sparrow	*Ciak Urasi*

Baya Weaver	*Ciak Tempua*
Chestnut Munia	*Pipit Rawa*
White-headed Munia	*Pipit Uban*
Scaly-breasted Munia	*Pipit Pinang*
White-rumped Munia	*Pipit Tuli*
White-bellied Munia	*Pipit Padi*
Pin-tailed Parrotfinch	*Ciak Ekor Jarum*
Hill Myna	*Tiong Mas*
Common Myna	*Tiong Gembala Kerbau*
White-vented Myna	*Tiong Tongkeng Putih*
Asian Glossy Starling	*Perling Mata Merah*
Purple-backed Starling	*Perling Belakang Ungu*
Black-naped Oriole	*Dendang Selayang*
Dark-throated Oriole	*Dendang Senja*
Black-and-crimson Oriole	*Dendang Mas*
Greater Racket-tailed Drongo	*Cecawi Anting-anting*
Lesser Racket-tailed Drongo	*Cecawi Hamba Kera*
Bronzed Drongo	*Cecawi Keladi*
Ashy Drongo	*Cecawi Rantau*
House Crow	*Gagak Rumah*
Large-billed Crow	*Gagak Paruh Besar*
Crested Jay	*Gagak Jerit*
Black Magpie	*Gagak Kambing*
Green Magpie	*Gagak Gunung*

Sources: Rahman, A. I. (1981), *Senarai Nama-nama Burung Semenanjung Malaysia dan Asia Tenggara*, Ibu Pejabat Perhilitan, Kuala Lumpur; G. W. H. Davison, Karen Phillipps, and Alias Kamis (1989), *Pengenalan Burung-burung Malaysia*, World Wide Fund for Nature Malaysia.

Index to Genera, Systematic Section

References in bold type refer to Colour Plates